Solutions Manual

Elementary Linear Algebra

C. H. Edwards, Jr.
David E. Penney
The University of Georgia, Athens

PRENTICE-HALL, Upper Saddle River, NJ 07458

Editorial/production supervision and
interior design: Sara Colacurto
Manufacturing buyer: Paula Benevento

A Pearson Education Company
Upper Saddle River, NJ 07458

Printed in the United States of America

13

ISBN 0-13-258278-3 01

Prentice-Hall International (UK) Limited,London
Prentice-Hall of Australia Pty. Limited, Sydney
Prentice-Hall Canada Inc., Toronto
Prentice-Hall Hispanoamericana, S.A., Mexico
Prentice-Hall of India Private Limited, New Delhi
Prentice-Hall of Japan, Inc., Tokyo
Pearson Education Asia Pte. Ltd., Singapore
Editora Prentice-Hall do Brasil, Ltda., Rio de Janeiro

CONTENTS

PREFACE

This is a solutions manual to accompany **ELEMENTARY LINEAR ALGEBRA** (1988) by C. H. Edwards, Jr. and David E. Penney. We provide here either the answer or a solution to essentially every problem in the text. In some cases only the answer is given to a problem that is entirely routine or involves only a standard elementary technique, but otherwise at least an outline of the principal steps in a complete solution is offered. In every case our goal is to provide just that amount of detail in the way of answers, hints, and suggestions that will supplement the text discussion most instructively. In most cases we believe that giving the highlights to an argument or computation, but leaving something for the student to do, is better than including a complete solution that students could simply transcribe. Many of the sections in this manual begin with comments on the priority of topics within the corresponding section of the text.

The purpose of this manual is assistance in the teaching and learning of the subject of elementary linear algebra. To this end we invite suggestions from those who use it (both professors and students) as to what features might be added or improved to increase its usefulness in future editions.

We are indebted to Alice F. Edwards for her expert preparation of the camera-ready copy for this manual.

C. H. E., Jr.
D. E. P.

CHAPTER 1

LINEAR SYSTEMS AND MATRICES

The first five sections of this chapter introduce fundamental concepts on which the remainder of the text is based, and therefore should be covered thoroughly in every course. The final three sections, however, treat optional applications that can be omitted or deferred as the instructor desires.

SECTION 1.1

INTRODUCTION TO LINEAR SYSTEMS

This initial section takes account of the fact that many students beginning elementary linear algebra remember only hazily the method of elimination for 2 by 2 and 3 by 3 systems. Moreover, high school algebra courses generally emphasize only the case in which a unique solution exists. Here we treat on an equal footing the other two cases -- in which either no solution exists or infinitely many solutions exist. Our geometric discussion of these three possibilites is based on a minimal knowledge of coordinate geometry: A linear equation $ax + by = c$ describes a straight line in the xy-plane, and any two distinct straight lines either are parallel or intersect in a single point. Lines and planes are treated in detail in Chapter 3.

The solutions to Problems 1-22 are quite routine and systematic, so we give only the answer. A numerical answer specifies a unique solution, while one involving the parameter t describes an infinite solution set.

1. $x = 3, \quad y = 2$

2. $x = 5, \quad y = -3$

3. $x = -4, \quad y = 3$

4. $x = 5, \quad y = 4$

5. No solution exists

6. No solution exists

7. $x = -10 + 4t, \quad y = t$

8. $x = 4 + 2t, \quad y = t$

9. $x = 4, \quad y = -1, \quad z = 3$

10. $x = 3, \quad y = 1, \quad z = -2$

11. $x = 1, \quad y = 3, \quad z = -4$

12. $x = 1, \quad y = 3, \quad z = 5$

13. $x = 0, \quad y = 0, \quad z = 0$

14. $x = 5, \quad y = 3, \quad z = -4$

15. No solution exists

16. No solution exists

17. No solution exists

18. No solution exists

19. $x = 8 + 3t, \quad y = 3 + 2t, \quad z = t$

20. $x = -5t, \quad y = 5 + t, \quad z = t$

21. $x = 3 - 2t, \quad y = 2 + 3t, \quad z = t$

22. $x = -4t, \quad y = -5t, \quad z = t$

23. The origin $(0,0)$ satisfies both equations, so the two lines represented by the two equations both pass through the origin. If the two lines coincide then there are infinitely many solutions. Otherwise $x = 0, y = 0$ is the unique solution.

24. (a) If the two planes are parallel then the system has no solution. Otherwise they either coincide or intersect in a line, and so the system has infinitely many solutions.
(b) If $d_1 = d_2 = 0$ then the origin $(0,0,0)$ lies in both planes, so part (a) implies that there are infinitely many solutions.

25. (a) No solution (b) A unique solution (c) No solution (d) No solution (e) A unique solution (f) Infinitely many solutions

26. (a) No solution (b) Infinitely many solutions (c) No solution (d) No solution (e) Infinitely many

8. solutions (f) A unique solution

SECTION 1.2

MATRICES AND GAUSSIAN ELIMINATION

This section describes Gaussian elimination (followed by back-substitution) as a systematic and practical method for the solution of systems of linear equations. Section 9.3 contains computer programs that students can use to practice the reduction of matrices to echelon form. After entering an arbitrary matrix (of size up to 10 by 10), the student is presented at each step with a menu asking him to specify the elementary row operation that he wants to perform next. At the press of a key it is done and the new matrix is displayed. The use of such programs can enable students to develop effective elimination tactics without getting bogged down in tedious arithmetic.

The linear systems in Problems 1-10 are already in echelon form, so we give only the answers to these problems.

1. $x_1 = 1, \quad x_2 = 0, \quad x_3 = 2$

2. $x_1 = 5, \quad x_2 = 1, \quad x_3 = -3$

3. $x_1 = 13 + 11t, \quad x_2 = 2 + 5t, \quad x_3 = t$

4. $x_1 = 35 + 33t, \quad x_2 = 5 + 7t, \quad x_3 = t$

5. $x_1 = 13 + 4t, \quad x_2 = 6 + t, \quad x_3 = 5 + 3t, \quad x_4 = t$

6. $x_1 = 17 + t, \quad x_2 = 11 + 3t, \quad x_3 = t, \quad x_4 = -4$

7. $x_1 = 3 - 8s + 19t, \quad x_2 = 7 + 2s - 7t, \quad x_3 = s, \quad x_4 = t$

8. $x_1 = -25 + 10s + 22t, \quad x_2 = 10 - 3t, \quad x_3 = s, \quad x_4 = t$

9. $x_1 = 1, \quad x_2 = 3, \quad x_3 = -5, \quad x_4 = 6$

10. $x_1 = 63s - 16t, \quad x_2 = 13s - 8t, \quad x_3 = s, \quad x_4 = 5t, \quad x_5 = t$

In each of Problems 11-22 we list first an echelon form E of the augmented coefficient matrix of the given linear system, and then the resulting solution (if any) of the system. The student should understand that the echelon matrix E is not unique, so a different sequence of elementary row operations may produce a different echelon matrix.

11. $$E = \begin{bmatrix} 1 & 3 & 2 & 5 \\ 0 & 1 & 0 & -2 \\ 0 & 0 & 1 & 4 \end{bmatrix}$$

$x_1 = 3, \quad x_2 = -2, \quad x_3 = 4$

12. $$E = \begin{bmatrix} 1 & -6 & -4 & 15 \\ 0 & 1 & 0 & -3 \\ 0 & 0 & 1 & 2 \end{bmatrix}$$

$x_1 = 5, \quad x_2 = -3, \quad x_3 = 2$

13. $$E = \begin{bmatrix} 1 & 3 & 3 & 13 \\ 0 & 1 & 2 & 3 \\ 0 & 0 & 0 & 0 \end{bmatrix}$$

$x_1 = 4 + 3t, \quad x_2 = 3 - 2t, \quad x_3 = t$

14. $$E = \begin{bmatrix} 1 & -2 & -2 & -9 \\ 0 & 0 & 1 & 7 \\ 0 & 0 & 0 & 0 \end{bmatrix}$$

$x_1 = 5 + 2t, \quad x_2 = t, \quad x_3 = 7$

15. $E = \begin{bmatrix} 1 & 1 & 1 & 1 \\ 0 & 1 & 3 & 3 \\ 0 & 0 & 0 & 1 \end{bmatrix}$

No solution

16. $E = \begin{bmatrix} 1 & -4 & -7 & 6 \\ 0 & 1 & 2 & 0 \\ 0 & 0 & 0 & 1 \end{bmatrix}$

No solution

17. $E = \begin{bmatrix} 1 & -4 & -3 & -3 & 4 \\ 0 & 1 & 0 & -1 & -4 \\ 0 & 0 & 1 & 3 & 5 \end{bmatrix}$

$x_1 = 3 - 2t, \quad x_2 = -4 + t, \quad x_3 = 5 - 3t, \quad x_4 = t$

18. $E = \begin{bmatrix} 1 & -2 & -4 & -13 & -8 \\ 0 & 0 & 1 & 4 & 3 \\ 0 & 0 & 0 & 0 & 0 \end{bmatrix}$

$x_1 = 4 + 2s - 3t, \quad x_2 = s, \quad x_3 = 3 - 4t, \quad x_4 = t$

19. $E = \begin{bmatrix} 1 & -2 & 5 & -5 & -7 \\ 0 & 1 & -2 & 3 & 5 \\ 0 & 0 & 0 & 0 & 0 \end{bmatrix}$

$x_1 = 3 - s - t, \quad x_2 = 5 + 2s - 3t, \quad x_3 = s, \quad x_4 = t$

20. $E = \begin{bmatrix} 1 & 3 & 2 & -7 & 3 & 9 \\ 0 & 1 & 3 & -7 & 2 & 7 \\ 0 & 0 & 1 & -2 & 0 & 2 \end{bmatrix}$

$x_1 = 2 + 3t,\ x_2 = 1 + s - 2t,\ x_3 = 2 + 2s,\ x_4 = s,\ x_5 = t$

21. $E = \begin{bmatrix} 1 & 1 & 1 & 0 & 6 \\ 0 & 1 & 5 & 1 & 20 \\ 0 & 0 & 1 & 0 & 3 \\ 0 & 0 & 0 & 1 & 4 \end{bmatrix}$

$x_1 = 2,\quad x_2 = 1,\quad x_3 = 3,\quad x_4 = 4$

22. $E = \begin{bmatrix} 1 & -2 & -4 & 0 & -9 \\ 0 & 1 & 6 & 1 & 21 \\ 0 & 0 & 1 & 0 & 4 \\ 0 & 0 & 0 & 1 & -1 \end{bmatrix}$

$x_1 = 3,\quad x_2 = -2,\quad x_3 = 4,\quad x_4 = -1$

23. An echelon form of the augmented matrix is

$$\begin{bmatrix} 3 & 2 & 1 \\ 0 & 0 & k-2 \end{bmatrix}$$

Hence there are infinitely many solutions if $k = 2$, but no solution if $k \neq 2$.

24. An echelon form of the augmented matrix is

$$\begin{bmatrix} 3 & 2 & 0 \\ 0 & k-4 & 0 \end{bmatrix}$$

Hence there are infinitely many solutions if $k = 4$, but only the unique solution $(0,0)$ if $k \neq 4$.

25. An echelon form of the augmented matrix is

$$\begin{bmatrix} 3 & 2 & 11 \\ 0 & k-4 & 1 \end{bmatrix}$$

Hence there is a unique solution if $k \neq 4$, but no solution if $k = 4$.

26. An echelon form of the augmented matrix is

$$\begin{bmatrix} 1 & 1 & k-2 \\ 0 & 1 & 3k-7 \end{bmatrix}$$

Hence for every value of k there is a unique solution.

27. An echelon form of the augmented matrix is

$$\begin{bmatrix} 1 & 2 & 1 & 3 \\ 0 & 5 & 5 & 1 \\ 0 & 0 & 0 & k-11 \end{bmatrix}$$

Hence there are infinitely many solutions if $k = 11$, but no solution if $k \neq 11$.

28. An echelon form of the augmented matrix is

$$\begin{bmatrix} 1 & 2 & 1 & b \\ 0 & -5 & 1 & a-2b \\ 0 & 0 & 0 & c-2a-3b \end{bmatrix}$$

Hence there are infinitely many solutions if $c = 2a + 3b$, but no solution otherwise.

30. (a) This part is essentially obvious, because a multiple of an equation that is satisfied is also satisfied, and the sum of two equations that are satisfied is one that is satisfied.

(b) Let us write $A_1 = B_1, B_2, \cdots, B_n, B_{n+1} = A_2$ where each B_{k+1} is obtained from B_k by a single elementary row operation $(k = 1, 2, \cdots, n)$. Then it follows by n applications of part (a) that every solution of the system S_1 associated with A_1 is also a solution of the system S_2 associated with A_2. But Problem 29(d) implies that A_1 also can be obtained from A_2 by elementary row operations, so by the same token every solution of S_2 is also a solution of S_1.

SECTION 1.3

GAUSS-JORDAN ELIMINATION

Our first theorems and proofs appear in the latter part of this section, which may seem a bit theoretical to some students ("What's so interesting about a trivial solution?"). Now's the time for students to find out that the study of linear algebra is a quest for comprehensive understanding of linear systems, not just a collection of solution methods. The importance of the basic theory of linear systems that begins to unfold here cannot be overemphasized; it will be applied repeatedly throughout the remainder of the text.

Each of the matrices in Problems 1-20 can be transformed to reduced echelon form without the appearance of any fractions. The main thing is to get started right. For instance, begin Problem 3 by subtracting the second row from the first row to get a leading 1. In Problem 4, first subtract the first row from the second row, and then subtract the (new) second row from the first row.

1. $\begin{bmatrix} 1 & 0 \\ 0 & 1 \end{bmatrix}$

2. $\begin{bmatrix} 1 & 0 \\ 0 & 1 \end{bmatrix}$

3. $\begin{bmatrix} 1 & 0 & -2 \\ 0 & 1 & 3 \end{bmatrix}$

4. $\begin{bmatrix} 1 & 0 & 2 \\ 0 & 1 & -1 \end{bmatrix}$

5. $\begin{bmatrix} 1 & 0 & -5 \\ 0 & 1 & -3 \end{bmatrix}$

6. $\begin{bmatrix} 1 & 0 & 7 \\ 0 & 1 & -6 \end{bmatrix}$

7. $\begin{bmatrix} 1 & 0 & 5 \\ 0 & 1 & -1 \\ 0 & 0 & 0 \end{bmatrix}$

8. $\begin{bmatrix} 1 & 0 & 0 \\ 0 & 1 & 0 \\ 0 & 0 & 1 \end{bmatrix}$

9. $\begin{bmatrix} 1 & 0 & 2 \\ 0 & 1 & 4 \\ 0 & 0 & 0 \end{bmatrix}$

10. $\begin{bmatrix} 1 & 0 & -3 \\ 0 & 1 & 5 \\ 0 & 0 & 0 \end{bmatrix}$

11. $\begin{bmatrix} 1 & 3 & 0 \\ 0 & 0 & 1 \\ 0 & 0 & 0 \end{bmatrix}$

12. $\begin{bmatrix} 1 & -4 & 0 \\ 0 & 0 & 1 \\ 0 & 0 & 0 \end{bmatrix}$

13. $\begin{bmatrix} 1 & 0 & 0 & 3 \\ 0 & 1 & 0 & -2 \\ 0 & 0 & 1 & 2 \end{bmatrix}$

14. $\begin{bmatrix} 1 & 0 & 0 & 4 \\ 0 & 1 & 0 & -3 \\ 0 & 0 & 1 & 5 \end{bmatrix}$

15. $\begin{bmatrix} 1 & 0 & -1 & 2 \\ 0 & 1 & 3 & -1 \\ 0 & 0 & 0 & 0 \end{bmatrix}$

16. $\begin{bmatrix} 1 & 0 & 3 & -2 \\ 0 & 1 & 4 & 3 \\ 0 & 0 & 0 & 0 \end{bmatrix}$

17. $\begin{bmatrix} 1 & 0 & 0 & 2 & -3 \\ 0 & 1 & 0 & -1 & 4 \\ 0 & 0 & 1 & -2 & -5 \end{bmatrix}$

18. $\begin{bmatrix} 1 & 0 & 3 & -2 & 3 \\ 0 & 1 & 4 & 5 & 1 \\ 0 & 0 & 0 & 0 & 0 \end{bmatrix}$

19. $\begin{bmatrix} 1 & 0 & 2 & 1 & 3 \\ 0 & 1 & -2 & -3 & 1 \\ 0 & 0 & 0 & 0 & 0 \end{bmatrix}$

20. $\begin{bmatrix} 1 & 2 & 0 & 2 & 3 \\ 0 & 0 & 1 & 1 & 4 \\ 0 & 0 & 0 & 0 & 0 \end{bmatrix}$

In each of Problems 21-30 we list first the <u>reduced</u> echelon form E of the augmented coefficient matrix of the given linear system, and then the resulting solution (if any) of the system.

21. $E = \begin{bmatrix} 1 & 0 & 0 & 3 \\ 0 & 1 & 0 & -2 \\ 0 & 0 & 1 & 4 \end{bmatrix}$

$x_1 = 3, \quad x_2 = -2, \quad x_3 = 4$

22. $E = \begin{bmatrix} 1 & 0 & 0 & 5 \\ 0 & 1 & 0 & -3 \\ 0 & 0 & 1 & 2 \end{bmatrix}$

$x_1 = 5, \quad x_2 = -3, \quad x_3 = 2$

23. $E = \begin{bmatrix} 1 & 0 & -3 & 4 \\ 0 & 1 & 2 & 3 \\ 0 & 0 & 0 & 0 \end{bmatrix}$

$x_1 = 4 + 3t, \quad x_2 = 3 - 2t, \quad x_3 = t$

24. $E = \begin{bmatrix} 1 & -2 & 0 & 5 \\ 0 & 0 & 1 & 7 \\ 0 & 0 & 0 & 0 \end{bmatrix}$

$x_1 = 5 + 2t, \quad x_2 = t, \quad x_3 = 7$

25. $E = \begin{bmatrix} 1 & 0 & -2 & 0 \\ 0 & 1 & 3 & 0 \\ 0 & 0 & 0 & 1 \end{bmatrix}$

No solution

26. $E = \begin{bmatrix} 1 & 0 & 1 & 0 \\ 0 & 1 & 2 & 0 \\ 0 & 0 & 0 & 1 \end{bmatrix}$

No solution

27. $E = \begin{bmatrix} 1 & 0 & 0 & 2 & 3 \\ 0 & 1 & 0 & -1 & -4 \\ 0 & 0 & 1 & 3 & 5 \end{bmatrix}$

$x_1 = 3 - 2t$, $x_2 = -4 + t$, $x_3 = 5 - 3t$, $x_4 = t$

28. $E = \begin{bmatrix} 1 & -2 & 0 & 3 & 4 \\ 0 & 0 & 1 & 4 & 3 \\ 0 & 0 & 0 & 0 & 0 \end{bmatrix}$

$x_1 = 4 + 2s - 3t$, $x_2 = s$, $x_3 = 3 - 4t$, $x_4 = t$

29. $E = \begin{bmatrix} 1 & 0 & 1 & 1 & 3 \\ 0 & 1 & -2 & 3 & 5 \\ 0 & 0 & 0 & 0 & 0 \end{bmatrix}$

$x_1 = 3 - s - t$, $x_2 = 5 + 2s - 3t$, $x_3 = s$, $x_4 = t$

30. $E = \begin{bmatrix} 1 & 0 & 0 & 0 & -3 & 2 \\ 0 & 1 & 0 & -1 & 2 & 1 \\ 0 & 0 & 1 & -2 & 0 & 2 \end{bmatrix}$

$x_1 = 2 + 3t$, $x_2 = 1 + s - 2t$, $x_3 = 2 + 2s$, $x_4 = s$, $x_5 = t$

31. Routine

32. If $ad - bc \neq 0$, then not both a and c can be zero. We may therefore assume that $a \neq 0$, so

$$\begin{bmatrix} a & b \\ c & d \end{bmatrix} \longrightarrow \begin{bmatrix} 1 & b/a \\ c & d \end{bmatrix}$$

$$\longrightarrow \begin{bmatrix} 1 & b/a \\ 0 & d - bc/a \end{bmatrix} \longrightarrow \begin{bmatrix} 1 & b/a \\ 0 & ad - bc \end{bmatrix}$$

Finally divide the second row by $ad - bc$, and then subtract b/a times the second row from the first row.

33. $\begin{bmatrix} 1 & 0 \\ 0 & 1 \end{bmatrix}$ $\begin{bmatrix} 1 & * \\ 0 & 0 \end{bmatrix}$ $\begin{bmatrix} 0 & 1 \\ 0 & 0 \end{bmatrix}$ $\begin{bmatrix} 0 & 0 \\ 0 & 0 \end{bmatrix}$

34. $\begin{bmatrix} 1 & 0 & 0 \\ 0 & 1 & 0 \\ 0 & 0 & 1 \end{bmatrix}$ $\begin{bmatrix} 1 & 0 & * \\ 0 & 1 & * \\ 0 & 0 & 0 \end{bmatrix}$ $\begin{bmatrix} 1 & * & 0 \\ 0 & 0 & 1 \\ 0 & 0 & 0 \end{bmatrix}$

$\begin{bmatrix} 1 & * & * \\ 0 & 0 & 0 \\ 0 & 0 & 0 \end{bmatrix}$ $\begin{bmatrix} 0 & 1 & * \\ 0 & 0 & 0 \\ 0 & 0 & 0 \end{bmatrix}$ $\begin{bmatrix} 0 & 0 & 1 \\ 0 & 0 & 0 \\ 0 & 0 & 0 \end{bmatrix}$ $\begin{bmatrix} 0 & 0 & 0 \\ 0 & 0 & 0 \\ 0 & 0 & 0 \end{bmatrix}$

35. (a) $a(kx_0) + b(ky_0) = k(ax_0 + by_0) = 0$

$c(kx_0) + d(ky_0) = k(cx_0 + dy_0) = 0$

(b) $a(x_1 + x_2) + b(y_1 + y_2) = (ax_1 + by_1) + (ax_2 + by_2) = 0$

$c(x_1 + x_2) + d(y_1 + y_2) = (cx_1 + dy_1) + (cx_2 + dy_2) = 0$

36. By Problem 32 the coefficient matrix is row-equivalent to the 2 by 2 identity matrix. Therefore Theorem 4 implies that the system has only the trivial solution.

37. If $ad - bc = 0$ then, much as in Problem 32, we see that the second row of the reduced echelon matrix is zero. Hence there is a free variable, and thus the system has a nontrivial solution.

38. By Problem 37 there is a nontrivial solution if and only if

$$(c + 2)(c - 3) - (2)(3) = c^2 - c - 12 = (c - 4)(c + 3) = 0,$$

that is, $c = 4$ or $c = -3$.

39. It is given that the augmented matrix is

$$\begin{bmatrix} a_{11} & a_{12} & a_{13} & 0 \\ a_{21} & a_{22} & a_{23} & 0 \\ pa_{11}+qa_{21} & pa_{12}+qa_{22} & pa_{13}+qa_{23} & 0 \end{bmatrix}.$$

Upon subtracting p times the first row and also q times the second row from the third row, we get

$$\begin{bmatrix} a_{11} & a_{12} & a_{13} & 0 \\ a_{21} & a_{22} & a_{23} & 0 \\ 0 & 0 & 0 & 0 \end{bmatrix}.$$

Hence there is at least one free variable, and so the system has a nontrivial solution.

40. In reducing further from the echelon matrix E to E^*, the leading entries of E become the leading ones in the reduced echelon matrix E^*. Thus the nonzero rows of E^* come precisely from the nonzero rows of E. Hence we are talking about the same rows in either case.

SECTION 1.4

MATRIX OPERATIONS

The objective of this section is simple to state: It is not merely knowledge of, but complete mastery of matrix addition and multiplication (particularly the latter). Matrix multiplication must be practised until it is carried out not only accurately but also quickly and with confidence -- until the student cannot look at two matrices A and B without thinking of "pouring" the ith row of A down the jth column of B.

1. $\begin{bmatrix} 5 & -15 \\ 18 & 5 \end{bmatrix}$

2. $\begin{bmatrix} 16 & -9 & -18 \\ -26 & 22 & 15 \end{bmatrix}$

3. $\begin{bmatrix} -26 & 20 \\ 12 & -6 \\ 22 & 18 \end{bmatrix}$

4. $\begin{bmatrix} 44 & -22 & -20 \\ 53 & 10 & -26 \\ 35 & 21 & 94 \end{bmatrix}$

5. $AB = \begin{bmatrix} -9 & 1 \\ -10 & 12 \end{bmatrix}$ $BA = \begin{bmatrix} -2 & 8 \\ 11 & 5 \end{bmatrix}$

6. $AB = \begin{bmatrix} 7 & -13 & -24 \\ 23 & 10 & 41 \\ 11 & -8 & 57 \end{bmatrix}$ $BA = \begin{bmatrix} 1 & -17 & -22 \\ 12 & 16 & 7 \\ 27 & -21 & 57 \end{bmatrix}$

7. $AB = [26]$ $BA = \begin{bmatrix} 3 & 6 & 9 \\ 4 & 8 & 12 \\ 5 & 10 & 15 \end{bmatrix}$

8. $AB = \begin{bmatrix} 21 & 15 \\ 35 & 0 \end{bmatrix}$ $BA = \begin{bmatrix} 3 & 0 & 9 \\ 7 & -20 & 13 \\ 16 & -25 & 38 \end{bmatrix}$

9. $BA = \begin{bmatrix} 4 \\ 7 \\ -22 \end{bmatrix}$ AB is undefined.

10. $AB = \begin{bmatrix} 1 & -2 & 13 \\ 5 & -6 & 31 \end{bmatrix}$ BA is undefined.

11. $AB = [11 \quad 1 \quad 5 \quad 3]$ BA is undefined.

12. Neither AB nor BA is defined.

13. $ABC = \begin{bmatrix} 32 & 51 \\ -2 & -17 \end{bmatrix}$

14. $ABC = [-3]$

15. $ABC = \begin{bmatrix} 12 & 15 \\ 8 & 10 \end{bmatrix}$

16. $ABC = \begin{bmatrix} -4 & -4 & -2 & 2 \\ -9 & -12 & -9 & 12 \\ -14 & -18 & -13 & 17 \end{bmatrix}$

Each of the linear systems in Problems 17-22 is already in echelon form, so it remains only (by back substitution) to write the solution, first in parametric form and then in vector form.

17. $x_3 = s, \quad x_4 = t, \quad x_1 = 5s - 4t, \quad x_2 = -2s + 7t$

$\bar{x} = s(5,-2,1,0) + t(-4,7,0,1)$

18. $x_2 = s, \quad x_4 = t, \quad x_1 = 3s - 6t, \quad x_3 = -9t$

$\bar{x} = s(3,1,0,0) + t(-6,0,-9,1)$

19. $x_4 = s, \quad x_5 = t, \quad x_1 = -3s + t, \quad x_2 = 2s - 6t, \quad x_3 = -s + 8t$

$\bar{x} = s(-3,2,-1,1,0) + t(1,-6,8,0,1)$

20. $x_2 = s,\ x_5 = t,\ x_1 = 3s - 7t,\ x_3 = 2t,\ x_4 = 10t$

$x = s(3,1,0,0,0) + t(-7,0,2,10,1)$

21. $x_3 = r,\ x_4 = s,\ x_5 = t,\ x_1 = r - 2s - 7t,\ x_2 = -2r + 3s - 4t$

$\bar{x} = r(1,-2,1,0,0) + s(-2,3,0,1,0) + t(-7,-4,0,0,1)$

22. $x_2 = r,\ x_4 = s,\ x_5 = t,\ x_1 = r - 7s - 3t,\ x_3 = s + 2t$

$\bar{x} = r(1,1,0,0,0) + s(-7,0,1,1,0) + t(-3,0,2,0,1)$

23. The four equations

$$2a + c = 1 \qquad 2b + d = 0$$
$$3a + 2c = 0 \qquad 3b + 2d = 1$$

are readily solved for $a = 2,\ b = -1,\ c = -3,\ d = 2.$ Hence

$$B = \begin{bmatrix} 2 & -1 \\ -3 & 2 \end{bmatrix}.$$

24. $B = \begin{bmatrix} 7 & -4 \\ -5 & 3 \end{bmatrix}$

25. $B = \begin{bmatrix} 3 & -7 \\ -2 & 5 \end{bmatrix}$

26. The four equations

$$a - 2c = 1 \qquad b - 2d = 0$$
$$-2a + 4c = 0 \qquad -2b + 4d = 1$$

are obviously inconsistent, so B does not exist.

27. Each diagonal element of AB is the product of the corresponding diagonal elements of A and B.

28. The point is simply that (by associativity) parentheses don't matter:

$$A^rA^s = (AA\cdots A)(AA\cdots A) = (AAA\cdots AA) = A^{r+s},$$

the product of $r + s$ copies of A in either case.

29. Compute both A^2 and $(a + d)A - (ad - bc)I$ in terms of a,b,c,d. Then note that they are equal.

30. $A^2 = 4A - 3I = \begin{bmatrix} 5 & 4 \\ 4 & 5 \end{bmatrix}$ $\quad A^3 = 4A^2 - 3A = \begin{bmatrix} 14 & 13 \\ 13 & 14 \end{bmatrix}$

$A^4 = 4A^3 - 3A^2 = \begin{bmatrix} 41 & 40 \\ 40 & 41 \end{bmatrix}$ $\quad A^5 = 4A^4 - 3A^3 = \begin{bmatrix} 122 & 121 \\ 121 & 122 \end{bmatrix}$

31. (a) $(A+B)(A-B) = \begin{bmatrix} -25 & -34 \\ -71 & -34 \end{bmatrix}$ $\quad A^2 - B^2 = \begin{bmatrix} -8 & -45 \\ -44 & -51 \end{bmatrix}$

(b) $(A + B)(A - B) = A(A - B) + B(A - B)$
$= A^2 - AB + BA - B^2 = A^2 - B^2$

32. (a) $(A + B)^2 = \begin{bmatrix} 5 & 52 \\ -13 & 96 \end{bmatrix}$ $\quad A^2 + 2AB + B^2 = \begin{bmatrix} 22 & 41 \\ 14 & 79 \end{bmatrix}$

33. $\begin{bmatrix} 1 & 0 \\ 0 & 1 \end{bmatrix}$ $\begin{bmatrix} -1 & 0 \\ 0 & -1 \end{bmatrix}$ $\begin{bmatrix} +1 & 0 \\ 0 & -1 \end{bmatrix}$ $\begin{bmatrix} -1 & 0 \\ 0 & +1 \end{bmatrix}$

34. $\begin{bmatrix} 1 & -1 \\ 1 & -1 \end{bmatrix}$

35. $\begin{bmatrix} 2 & -1 \\ 2 & -1 \end{bmatrix}$

36. $\begin{bmatrix} 0 & 1 \\ 1 & 0 \end{bmatrix}$

37. $\begin{bmatrix} 0 & 1 \\ -1 & 0 \end{bmatrix}$

38. Let A and B be the two matrices listed in Problems 36 and 37. Then $A^2 + B^2 = I - I = 0$.

39. $A(c_1\bar{x}_1 + c_2\bar{x}_2) = c_1A\bar{x}_1 + c_2A\bar{x}_2 = c_1\bar{0} + c_2\bar{0} = \bar{0}$

40. (a) $A(\bar{x}_0 + \bar{x}_1) = A\bar{x}_0 + A\bar{x}_1 = \bar{0} + \bar{b} = \bar{b}$

(b) $A(\bar{x}_1 - \bar{x}_2) = A\bar{x}_1 - A\bar{x}_2 = \bar{b} - \bar{b} = \bar{0}$

41. $$\begin{aligned}(A + B)^3 &= (A + B)(A + B)^2 \\ &= (A + B)(A^2 + 2AB + B^2) \\ &= A(A^2 + 2AB + B^2) + B(A^2 + 2AB + B^2) \\ &= A^3 + 2A^2B + AB^2 \\ &\quad + A^2B + 2AB^2 + B^3\end{aligned}$$

$(A + B)^3 = A^3 + 3A^2B + 3AB^2 + B^3$

To compute $(A + B)^4$, write $(A + B)^4 = (A + B)(A + B)^3$ and proceed similarly.

42. (a) $N^2 = \begin{bmatrix} 0 & 0 & 4 \\ 0 & 0 & 0 \\ 0 & 0 & 0 \end{bmatrix}$ $N^3 = \begin{bmatrix} 0 & 0 & 0 \\ 0 & 0 & 0 \\ 0 & 0 & 0 \end{bmatrix}$

(b) $A^2 = \begin{bmatrix} 1 & 4 & 4 \\ 0 & 1 & 4 \\ 0 & 0 & 1 \end{bmatrix}$ $A^3 = \begin{bmatrix} 1 & 6 & 12 \\ 0 & 1 & 6 \\ 0 & 0 & 1 \end{bmatrix}$

$A^4 = \begin{bmatrix} 1 & 8 & 24 \\ 0 & 1 & 8 \\ 0 & 0 & 1 \end{bmatrix}$

43. $A^2 = \begin{bmatrix} 6 & -3 & -3 \\ -3 & 6 & -3 \\ -3 & -3 & 6 \end{bmatrix} = 3A$

Then $A^3 = 3A^2 = 3(3A) = 9A$, $A^4 = 3A^3 = 3(9A) = 27A$, and so forth.

SECTION 1.5

INVERSE MATRICES

The computational objective of this section is clearcut: Find the inverse of a given invertible matrix. From a theoretical viewpoint, Theorem 7 on the properties of nonsingular matrices summarizes most of the basic theory of this chapter. Hence a careful survey of all that goes into its proof is a good review of the theoretical knowledge that the student needs to carry forward from Chapter 1 to the remainder of the text.

1. $A^{-1} = \begin{bmatrix} 3 & -2 \\ -4 & 3 \end{bmatrix}$, $\bar{x} = \begin{bmatrix} 3 \\ -2 \end{bmatrix}$

2. $A^{-1} = \begin{bmatrix} 5 & -7 \\ -2 & 3 \end{bmatrix}$, $\bar{x} = \begin{bmatrix} -26 \\ 11 \end{bmatrix}$

3. $A^{-1} = \begin{bmatrix} 6 & -7 \\ -5 & 6 \end{bmatrix}$, $\bar{x} = \begin{bmatrix} 33 \\ -28 \end{bmatrix}$

4. $A^{-1} = \begin{bmatrix} 17 & -12 \\ -7 & 5 \end{bmatrix}$, $\bar{x} = \begin{bmatrix} 25 \\ -10 \end{bmatrix}$

5. $A^{-1} = \frac{1}{2}\begin{bmatrix} 4 & -2 \\ -5 & 3 \end{bmatrix}$, $\bar{x} = \frac{1}{2}\begin{bmatrix} 8 \\ -7 \end{bmatrix}$

6. $A^{-1} = \frac{1}{3}\begin{bmatrix} 6 & -7 \\ -3 & 4 \end{bmatrix}$, $\bar{x} = \frac{1}{3}\begin{bmatrix} 25 \\ -10 \end{bmatrix}$

7. $A^{-1} = \frac{1}{4}\begin{bmatrix} 7 & -9 \\ -5 & 7 \end{bmatrix}$, $\bar{x} = \frac{1}{4}\begin{bmatrix} 3 \\ -1 \end{bmatrix}$

8. $A^{-1} = \frac{1}{5}\begin{bmatrix} 10 & -15 \\ -5 & 8 \end{bmatrix}$, $\bar{x} = \frac{1}{5}\begin{bmatrix} 25 \\ -11 \end{bmatrix}$

9. $\begin{bmatrix} 5 & -6 \\ -4 & 5 \end{bmatrix}$

10. $\frac{1}{2}\begin{bmatrix} 6 & -7 \\ -4 & 5 \end{bmatrix}$

11. $\begin{bmatrix} -5 & -2 & 5 \\ 2 & 1 & -2 \\ -4 & -3 & 5 \end{bmatrix}$

12. $\begin{bmatrix} 18 & 2 & -7 \\ -3 & 0 & 1 \\ -4 & -1 & 2 \end{bmatrix}$

13. $\begin{bmatrix} -13 & 42 & -5 \\ 3 & -9 & 1 \\ 2 & -7 & 1 \end{bmatrix}$

14. $\begin{bmatrix} 11 & -7 & -9 \\ -4 & 3 & 3 \\ -2 & 1 & 2 \end{bmatrix}$

15. $\begin{bmatrix} -22 & 2 & 7 \\ -27 & 3 & 8 \\ 10 & -1 & -3 \end{bmatrix}$

16. $\frac{1}{3}\begin{bmatrix} -3 & 0 & 3 \\ -1 & -3 & -1 \\ -1 & 3 & 2 \end{bmatrix}$

17. $\frac{1}{4}\begin{bmatrix} -2 & -6 & -3 \\ -2 & -2 & -1 \\ -2 & -2 & 1 \end{bmatrix}$

18. $\frac{1}{5}\begin{bmatrix} 1 & 2 & -2 \\ -5 & 0 & 5 \\ -3 & -1 & 6 \end{bmatrix}$

19. $\frac{1}{6}\begin{bmatrix} -21 & 11 & 8 \\ 9 & -5 & -2 \\ -3 & 3 & 0 \end{bmatrix}$

20. $\frac{1}{7}\begin{bmatrix} 3 & 1 & 0 \\ -2 & -3 & 7 \\ -1 & 2 & 0 \end{bmatrix}$

21. $\begin{bmatrix} 0 & 1 & 0 & 0 \\ -2 & 0 & 1 & 0 \\ 1 & 0 & 0 & 0 \\ 0 & -3 & 0 & 1 \end{bmatrix}$

22. $\begin{bmatrix} 1 & -1 & 1 & 0 \\ 0 & -2 & -1 & 2 \\ 0 & 1 & 1 & -1 \\ -3 & 3 & -5 & 1 \end{bmatrix}$

23. $A^{-1} = \begin{bmatrix} 4 & -3 \\ -5 & 4 \end{bmatrix}$, $X = \begin{bmatrix} 7 & 18 & -35 \\ -9 & -23 & 45 \end{bmatrix}$

24. $A^{-1} = \begin{bmatrix} 7 & -6 \\ -8 & 7 \end{bmatrix}$, $X = \begin{bmatrix} 14 & -30 & 46 \\ -16 & 35 & -53 \end{bmatrix}$

25. $A^{-1} = \begin{bmatrix} 11 & -9 & 4 \\ -2 & 2 & -1 \\ -2 & 1 & 0 \end{bmatrix}$, $X = \begin{bmatrix} 7 & -14 & 15 \\ -1 & 3 & -2 \\ -2 & 2 & -4 \end{bmatrix}$

26. $A^{-1} = \begin{bmatrix} -16 & 3 & 11 \\ 6 & -1 & -4 \\ -13 & 2 & 9 \end{bmatrix}$, $X = \begin{bmatrix} -21 & 9 & 6 \\ 8 & -3 & -2 \\ -17 & 6 & 5 \end{bmatrix}$

27. $A^{-1} = \begin{bmatrix} 7 & -20 & 17 \\ 0 & -1 & 1 \\ -2 & 6 & -5 \end{bmatrix}$, $X = \begin{bmatrix} 17 & -20 & 24 & -13 \\ 1 & -1 & 1 & -1 \\ -5 & 6 & -7 & 4 \end{bmatrix}$

28. $A^{-1} = \begin{bmatrix} -2 & 2 & 1 \\ -4 & 3 & 3 \\ 11 & -9 & -7 \end{bmatrix}$, $X = \begin{bmatrix} -5 & 5 & 10 & 1 \\ -8 & 8 & 15 & 7 \\ 24 & -23 & -45 & -13 \end{bmatrix}$

29. (a) The fact that A^{-1} is the inverse of A means that $AA^{-1} = A^{-1}A = I$. Thus when A^{-1} is multiplied either on the right or on the left by A, the result is I. This means that A is the inverse of A^{-1}.

(b) $A^n(A^{-1})^n = A^{n-1}AA^{-1}(A^{-1})^{n-1}$

$= A^{n-1}I(A^{-1})^{n-1} = \cdots = I$

Similarly $(A^{-1})^nA^n = I$, so $(A^{-1})^n$ is the inverse of A^n.

30. $ABC \cdot C^{-1}B^{-1}A^{-1} = ABIB^{-1}A^{-1} = AIA^{-1} = I$, and similarly $C^{-1}B^{-1}A^{-1} \cdot ABC = I$.

31. Let $p = -r > 0$, $q = -s > 0$, and $B = A^{-1}$. Then

$$A^rA^s = A^{-p}A^{-q} = (A^{-1})^p(A^{-1})^q$$

$$= B^pB^q = B^{p+q} \quad \text{(positive exponents)}$$

$$= (A^{-1})^{p+q} = A^{-p-q} = A^{r+s}$$

as desired, and similarly

$$(A^r)^s = (A^{-p})^{-q} = (B^p)^{-q} = B^{-pq} = A^{pq} = A^{rs}.$$

32. Multiplication of $AB = AC$ on the left by A^{-1} yields $B = C$.

33. In particular $A\bar{e}_j = \bar{e}_j$ where $\bar{e}_j$ is the jth column of I. Hence it follows from Fact 2 that $AI = I$, so $A = I^{-1} = I$.

34. The invertibility of a diagonal matrix with nonzero diagonal elements follows immediately from the rule for multiplying diagonal matrices (Problem 27 in Section 1.4). The inverse of such a diagonal matrix is gotten by inverting each diagonal element.

35. If the jth column of A is all zeros and B is any n by n matrix, then the jth column of BA is all zeros, so $BA \neq I$. Hence A has no inverse matrix. Similarly, if the ith row of A is all zeros, then so is the ith row of AB.

36. If $ad - bc = 0$, then it follows easily that one row of A is a multiple of the other. Hence the reduced row echelon form of A is of the form

$$\begin{bmatrix} * & * \\ 0 & 0 \end{bmatrix}$$

rather than the 2 by 2 identity matrix. Therefore A is not invertible.

37. Direct multiplication shows that $AA^{-1} = A^{-1}A = I$.

38. $EA = \begin{bmatrix} 3 & 0 \\ 0 & 1 \end{bmatrix} \begin{bmatrix} a & b \\ c & d \end{bmatrix} = \begin{bmatrix} 3a & 3b \\ c & d \end{bmatrix}$

39. $EA = \begin{bmatrix} 1 & 0 & 0 \\ 0 & 1 & 0 \\ 2 & 0 & 1 \end{bmatrix} \begin{bmatrix} a_{11} & a_{12} & a_{13} \\ a_{21} & a_{22} & a_{23} \\ a_{31} & a_{32} & a_{33} \end{bmatrix}$

$$= \begin{bmatrix} a_{11} & a_{12} & a_{13} \\ a_{21} & a_{22} & a_{23} \\ a_{31} + 2a_{11} & a_{32} + 2a_{12} & a_{33} + 2a_{13} \end{bmatrix}$$

40. $EA = \begin{bmatrix} 0 & 1 & 0 \\ 1 & 0 & 0 \\ 0 & 0 & 1 \end{bmatrix} \begin{bmatrix} a_{11} & a_{12} & a_{13} \\ a_{21} & a_{22} & a_{23} \\ a_{31} & a_{31} & a_{33} \end{bmatrix} = \begin{bmatrix} a_{21} & a_{22} & a_{23} \\ a_{11} & a_{12} & a_{13} \\ a_{31} & a_{32} & a_{33} \end{bmatrix}$

41. This follows immediately from the fact that the ijth element of AB is the product of the ith row of A and the jth column of B.

42. Let $\bar{e}_i$ denote the ith <u>row</u> of I. Then $\bar{e}_i B = B_i$, the ith row of B. Hence the result in Problem 41 yields

$$IB = \begin{bmatrix} \bar{e}_1 \\ \bar{e}_2 \\ \vdots \\ \bar{e}_m \end{bmatrix} B = \begin{bmatrix} \bar{e}_1 B \\ \bar{e}_2 B \\ \vdots \\ \bar{e}_m B \end{bmatrix} = \begin{bmatrix} B_1 \\ B_2 \\ \vdots \\ B_m \end{bmatrix} = B$$

43. Let $E_1, E_2, \cdots, E_k$ be the elementary matrices corresponding to the elementary row operations that reduce A to B. Then Theorem 5 gives

$$B = E_k E_{k-1} \cdots E_2 E_1 A = GA$$

where $G = E_k E_{k-1} \cdots E_2 E_1$.

44. This follows immediately from the result in Problem 43, because an invertible matrix is row-equivalent to the identity matrix.

45. One can simply photocopy the portion of the proof of Theorem

7 that follows Equation (21). Starting only with the assumption that A and B are square matrices with $AB = I$, it is proved there that A and B are invertible.

46. If $C = AB$ is invertible, so C^{-1} exists, then

$$A(BC^{-1}) = I \quad \text{and} \quad (C^{-1}A)B = I.$$

Hence the fact that A and B are invertible follows immediately from Problem 45.

SECTION 1.6

LINEAR SYSTEMS AND CURVE FITTING

This is the first of three brief sections that illustrate applications of the material on linear systems and matrices in Sections 1.1 through 1.5. Although Section 1.6 is optional, it can serve as useful background to the discussion of least squares solutions in Chapter 5.

In Problems 1 and 2 only two data points are given, so we are looking for a straight line with equation $y = A + Bx$. Substitution of the coordinates of the two points into this equation gives two linear equations that we can solve for A and B.

1. $y = -2 + 3x$

2. $y = 4 - 7x$

Three data points are given in each of Problems 3 through 6, so we are looking for a quadratic polynomial of the form $y = A + Bx + Cx^2$. Substitution of the coordinates of the three points into this equation gives a system of three linear equations that we can solve for A, B, and C.

3. $y = 3 - 2x^2$

4. $y = (4x + 3x^2 - 4x^3)/3$

5. $y = 5 - 3x + x^2$

6. $y = -10 - 7x + 2x^2$

In each of Problems 7 through 10, four data points are given, so we are looking for a cubic polynomial as in Example 1.

7. $y = 2x - x^3$

8. $y = 5 + 3x - x^3$

9. $y = 4 + 3x + 2x^2 + x^3$

10. $y = 17 - 5x + 3x^2 - 2x^3$

In each of Problems 11-14 we fit the circle

$$x^2 + y^2 + Ax + By + C = 0$$

to the three given points, just as in Example 2.

11. $x^2 + y^2 - 6x - 4y = 12$; center (3,2) and radius 5

12. $x^2 + y^2 + 6x - 8y = 75$; center (-3,4) and radius 10

13. $x^2 + y^2 + 4x + 4y = 5$; center (-2,-2) and radius $\sqrt{13}$

14. $x^2 + y^2 - 10x - 24y = 0$; center (5,12) and radius 13

Problems 15 and 16 follow precisely the pattern of solution illustrated in Example 3.

15. $2x^2 + xy + y^2 = 10$

16. $5x^2 + 4xy + 2y^2 = 100$

17. When we substitute $y(1) = 5$ and $y(2) = 4$ in $y = A + B/x$, we get the equations

$$\begin{aligned} A + B &= 5 \\ A + 0.5B &= 4 \end{aligned}$$

that are readily solved for $A = 3$, $B = 2$. Hence the desired curve is $y = 3 + 2/x$.

18. When we substitute $y(1) = 2$, $y(2) = 20$, and $y(4) = 41$ in $y = Ax + B/x + C/x^2$, we get the equations

$$\begin{aligned} A + B + C &= 2 \\ 2A + B/2 + C/4 &= 20 \\ 4A + B/4 + C/16 &= 41 \end{aligned}$$

that are readily solved for $A = 10$, $B = 8$, and $C = -16$. Hence the desired curve is $y = 10x + 8/x - 16/x^2$.

SECTION 1.7

MATRICES AND POPULATION MODELS

This is another optional applied section, but it will provide good background and motivation for Chapter 6. Therefore, if Section 1.7 is not covered at this point in the course, it might profitably be inserted just prior to Chapter 6 to indicate why we are interested in powers of matrices.

For each problem we list the values of A^n and of the two populations when n is sufficiently large. In each case the situation is clear by the time we get to $n = 50$.

1. $A^n = \begin{bmatrix} 0.500 & 0.500 \\ 0.500 & 0.500 \end{bmatrix}$ $\quad C_n = S_n = 500$

2. $A^n = \begin{bmatrix} 0.250 & 0.250 \\ 0.750 & 0.750 \end{bmatrix}$ $\quad C_n = 250, \quad S_n = 750$

3. $A^n = \begin{bmatrix} 0.375 & 0.375 \\ 0.625 & 0.625 \end{bmatrix}$ $\quad C_n = 375, \quad S_n = 625$

4. $A^n = \begin{bmatrix} 0.333 & 0.333 \\ 0.667 & 0.667 \end{bmatrix}$ $\quad C_n = 333, \quad S_n = 667$

5. Same answers as in Problem 4

6. $A^n = \begin{bmatrix} 0.429 & 0.429 \\ 0.571 & 0.571 \end{bmatrix} \approx \frac{1}{7}\begin{bmatrix} 3 & 3 \\ 4 & 4 \end{bmatrix}$

$C_n = 429$, $S_n = 571$

7. $A^n = \begin{bmatrix} 0.000 & 0.000 \\ 0.000 & 0.000. \end{bmatrix}$ $F_n = R_n = 0$ (extinction)

8. $A^{50} = \begin{bmatrix} -32.338 & 115.806 \\ -77.204 & 276.479 \end{bmatrix}$

$F_{50} = 8347$ and $R_{50} = 19928$; both populations are increasing rapidly (population explosion).

9. $A^n = \begin{bmatrix} -1.000 & 2.500 \\ -0.800 & 2.000 \end{bmatrix}$ $F_n = 50$, $R_n = 40$

10. $A^n = \begin{bmatrix} 0.000 & 0.000 \\ 0.000 & 0.000 \end{bmatrix}$ $F_n = R_n = 0$ (extinction)

SECTION 1.8

MATRICES AND CRYPTOGRAPHY

This section is truly optional, and purely for fun. The answers below are obtained precisely as in the 2 by 2 example in the text.

1. 50, 31, 13, 47, 31, 23, 8, 29

2. 107, 33, 88, 42, 16, 34

3. 29, 84, 96, 93, 15, 51, 58, 59

4. 54, 16, 61, 65, 96, 27, 103, 110

5. 99, 99, 89, 170, 71, 68, 66, 123

6. 76, 32, 131, 91, 40, 177, 50, 21, 81, 57, 25, 112

7. VIOLIN

8. IRREGULARS

9. ELEMENTARY

10. IRENE ADLER (who for Sherlock Holmes always remained <u>the</u> woman; see "A Scandal in Bohemia")

11. CONAN DOYLE

12. BAKER STREET

CHAPTER 2

DETERMINANTS

SECTION 2.1

2 BY 2 DETERMINANTS

This first section of Chapter 2 includes a more complete treatment of 2 x 2 determinants than is usual. The formal properties of determinants are easy to understand (and establish) in the 2 x 2 case, and this background should help students when they proceed to n x n determinants in Section 2.2. Indeed, much of the remainder of Chapter 2 consists of generalizing the content of Section 2.1 from the case of 2 x 2 determinants to higher order determinants.

1. 0
2. 1
3. 100
4. -77
5. -100
6. -2
7. $x^2 + x$
8. $-y^2$
9. $(x - y)^2$
10. -2

In Problems 11-20 we give the value of the coefficient determinant Δ as well as the solution (x,y).

11. $\Delta = 1$, $x = 10$, $y = -7$

12. $\Delta = 1$, $x = -1$, $y = 1$

13. $\Delta = 1$, $x = 2$, $y = -4$

14. $\Delta = 1$, $x = 5$, $y = -3$

15. $\Delta = 2, \quad x = 6, \quad y = -3$

16. $\Delta = -2, \quad x = 1/2, \quad y = 0$

17. $\Delta = 3, \quad x = 2, \quad y = -3$

18. $\Delta = 13, \quad x = 23/13, \quad y = 2/13$

19. $\Delta = 17, \quad x = 1/17, \quad y = -13/17$

20. $\Delta = 29, \quad x = 5, \quad y = 0$

21. If B is obtained by multiplying the second column of A by k, then

$$\det B = \begin{vmatrix} a & kb \\ c & kd \end{vmatrix}$$

$$= a(kd) - (kb)c = k(ad - bc)$$

$$= k\begin{vmatrix} a & b \\ c & d \end{vmatrix} = k \det A$$

22. If B is obtained by interchanging the columns of A, then

$$\det B = \begin{vmatrix} b & a \\ d & c \end{vmatrix}$$

$$= bc - ad = -(ad - bc)$$

$$= -\begin{vmatrix} a & b \\ c & d \end{vmatrix} = -\det A.$$

23. If the two columns of A are identical, then

$$\det A = \begin{vmatrix} a & a \\ c & c \end{vmatrix} = ac - ac = 0$$

24. If the second column of B is the sum of the second columns of A_1 and A_2, then

$$\det B = \begin{vmatrix} a & b_1 + b_2 \\ c & d_1 + d_2 \end{vmatrix}$$

$$= a(d_1 + d_2) - (b_1 + b_2)c = (ad_1 - b_1c) + (ad_2 - b_2c)$$

$$= \begin{vmatrix} a & b_1 \\ c & d_1 \end{vmatrix} + \begin{vmatrix} a & b_2 \\ c & d_2 \end{vmatrix}$$

$\det B = \det A_1 + \det A_2$.

25. If B is obtained by adding k times the first column of A to the second column, then

$$\det B = \begin{vmatrix} a & b+ka \\ c & d+kc \end{vmatrix}$$

$$= \begin{vmatrix} a & b \\ c & d \end{vmatrix} + k\begin{vmatrix} a & a \\ c & c \end{vmatrix}$$

$$= \det A + k\cdot 0 = \det A.$$

26. $$\begin{vmatrix} a_{11} & b_1 \\ a_{21} & b_2 \end{vmatrix} = \begin{vmatrix} a_{11} & a_{11}x + a_{12}y \\ a_{21} & a_{21}x + a_{22}y \end{vmatrix}$$

$$= x\begin{vmatrix} a_{11} & a_{11} \\ a_{21} & a_{21} \end{vmatrix} + y\begin{vmatrix} a_{11} & a_{12} \\ a_{21} & a_{22} \end{vmatrix}$$

$$= x\cdot 0 + y\cdot \det A$$

$$\begin{vmatrix} a_{11} & b_1 \\ a_{21} & b_2 \end{vmatrix} = y\cdot \det A$$

Now division by $\det A \neq 0$ gives the desired formula.

27. $x = \begin{vmatrix} u & 8 \\ v & 5 \end{vmatrix} = 5u - 8v, \qquad y = \begin{vmatrix} 5 & u \\ 3 & v \end{vmatrix} = -3u + 5v$

28. The coefficient determinant is $\cos^2\theta + \sin^2\theta = 1$, so

$$x = \begin{vmatrix} u & -\sin\theta \\ v & \cos\theta \end{vmatrix} = u\cos\theta + v\sin\theta,$$

$$y = \begin{vmatrix} \cos\theta & u \\ \sin\theta & v \end{vmatrix} = -u\sin\theta + v\cos\theta.$$

29. We write the system in the form

$$\begin{aligned} (2-\lambda)x - \qquad\quad 3y &= 0 \\ 3x + (-2-\lambda)y &= 0, \end{aligned}$$

and then note that the coefficient determinant is

$$(2-\lambda)(-2-\lambda) + 9 = \lambda^2 + 5 \neq 0.$$

Hence our homogeneous system has only the unique trivial solution.

30. $\det\,[a_1x_1 + a_2\mathbf{x}_2 \quad y] = \det[a_1\mathbf{x}_1 \quad y] + \det[a_2\mathbf{x}_2 \quad y]$

$$= a_1\det[\mathbf{x}_1 \quad y] + a_2\det[\mathbf{x}_2 \quad y]$$

31. The ijth element of $(AB)^T$ is the jith element of AB, and hence is the product of the jth row of A and the ith column of B. The ijth element of B^TA^T is the product of the ith row of B^T and the jth column of A^T. Because transposition of a matrix changes the ith row to the ith column and vice versa, it follows that the ijth element of B^TA^T is the product of the jth row of A and the ith column of B. Thus the matrices $(AB)^T$ and B^TA^T have the same ijth elements, and so are equal.

32. $\det AB = \begin{vmatrix} a\bar{x} + b\bar{y} \\ c\bar{x} + d\bar{y} \end{vmatrix}$

$$= \begin{vmatrix} a\bar{x} \\ c\bar{x} + d\bar{y} \end{vmatrix} + \begin{vmatrix} b\bar{y} \\ c\bar{x} + d\bar{y} \end{vmatrix}$$

$$= ac\begin{vmatrix} \bar{x} \\ \bar{x} \end{vmatrix} + ad\begin{vmatrix} \bar{x} \\ \bar{y} \end{vmatrix} + bc\begin{vmatrix} \bar{y} \\ \bar{x} \end{vmatrix} + bd\begin{vmatrix} \bar{y} \\ \bar{y} \end{vmatrix}$$

$$= ad\begin{vmatrix} \bar{x} \\ \bar{y} \end{vmatrix} + bc\begin{vmatrix} \bar{y} \\ \bar{x} \end{vmatrix}$$

$$= (ad - bc)\begin{vmatrix} \bar{x} \\ \bar{y} \end{vmatrix}$$

det AB = det A det B

SECTION 2.2

HIGHER ORDER DETERMINANTS

The object of this section is the evaluation of determinants using row and column cofactor expansions. Students should learn to look always for a particularly advantageous row or column to use. If no row or column contains a zero, then a preliminary simplification by means of an elementary row or column operation is in order. The properties of higher order determinants are introduced here from a computational point of view, and theoretical applications are deferred to Sections 2.3 and 2.4.

In discussing Problems 1-24 we denote by A the matrix whose determinant is to be evaluated.

1. det A = 0 because A has two identical rows.

2. det A = 0 because A has an all zero column.

3. det A = -1 because A is obtained by interchanging the first two rows of the identity matrix.

4. det A = -1 because A is obtained by interchanging the first and third rows of the identity matrix.

5. det A = (1)(4)(6) = 24 because A is an upper triangular matrix.

matrix.

7. Expansion along the first row yields

$$\begin{vmatrix} 0 & -3 & 0 \\ 7 & 11 & 4 \\ 5 & -2 & 3 \end{vmatrix} = -(-3)\begin{vmatrix} 7 & 4 \\ 5 & 3 \end{vmatrix} = 3$$

8. Expansion along the second column yields

$$\begin{vmatrix} 5 & 0 & 6 \\ 9 & -3 & -1 \\ 4 & 0 & 5 \end{vmatrix} = +(-3)\begin{vmatrix} 5 & 6 \\ 4 & 5 \end{vmatrix} = -3$$

9. Expansion along the second column yields

$$\begin{vmatrix} 4 & 0 & 6 \\ 5 & 0 & 8 \\ 7 & -4 & -9 \end{vmatrix} = -(-4)\begin{vmatrix} 4 & 6 \\ 5 & 8 \end{vmatrix} = 8$$

10. Expansion along the third row yields

$$\begin{vmatrix} 2 & 4 & 5 \\ -1 & 3 & 6 \\ 0 & 0 & 7 \end{vmatrix} = +(7)\begin{vmatrix} 2 & 4 \\ -1 & 3 \end{vmatrix} = 70$$

11. Expansion along the first column yields

$$\begin{vmatrix} 0 & 2 & -3 \\ 1 & 4 & -2 \\ 3 & 5 & 2 \end{vmatrix} = -(1)\begin{vmatrix} 2 & -3 \\ 5 & 2 \end{vmatrix} +(3)\begin{vmatrix} 2 & -3 \\ 4 & -2 \end{vmatrix} = 5$$

12. Expansion along the third column gives

$$\begin{vmatrix} 2 & -1 & 3 \\ 3 & -4 & 0 \\ -1 & 2 & 1 \end{vmatrix} = +(3)\begin{vmatrix} 3 & -4 \\ -1 & 2 \end{vmatrix} + (1)\begin{vmatrix} 2 & -1 \\ 3 & -4 \end{vmatrix} = 1$$

13. $\begin{vmatrix} 0 & 0 & 3 \\ 4 & 0 & 0 \\ 0 & 5 & 0 \end{vmatrix} = 3\begin{vmatrix} 4 & 0 \\ 0 & 5 \end{vmatrix} = 60$

14. $\begin{vmatrix} 2 & 1 & 0 \\ 1 & 2 & 1 \\ 0 & 1 & 2 \end{vmatrix} = (2)\begin{vmatrix} 2 & 1 \\ 1 & 2 \end{vmatrix} -(1)\begin{vmatrix} 1 & 1 \\ 0 & 2 \end{vmatrix} = 4$

15. Expansion along the first row yields

$$\det A = (1)\begin{vmatrix} 0 & 5 & 0 \\ 6 & 9 & 8 \\ 0 & 10 & 7 \end{vmatrix} = (1)(-5)\begin{vmatrix} 6 & 8 \\ 0 & 7 \end{vmatrix} = -210$$

16. Expansion along the third row yields

$$\det A = -(-3)\begin{vmatrix} 5 & 11 & 8 \\ 3 & -2 & 6 \\ 0 & 4 & 0 \end{vmatrix} = (3)(-4)\begin{vmatrix} 5 & 8 \\ 3 & 6 \end{vmatrix} = -72$$

17. Expansion along the first row and then along the second row yields

$$\det A = (+1)(+2)\begin{vmatrix} 0 & 3 & 0 \\ 0 & 0 & 4 \\ 5 & 0 & 0 \end{vmatrix} = (2)(-3)\begin{vmatrix} 0 & 4 \\ 5 & 0 \end{vmatrix} = 120$$

18. $$\det A = (5)\begin{vmatrix} 3 & 0 & -5 & 0 \\ -2 & 4 & 6 & 5 \\ 7 & 6 & 17 & 0 \\ 0 & 0 & 2 & 0 \end{vmatrix} = (5)(-2)\begin{vmatrix} 3 & 0 & 0 \\ -2 & 4 & 5 \\ 7 & 6 & 7 \end{vmatrix}$$

$$= (5)(-2)(3)\begin{vmatrix} 4 & 5 \\ 6 & 7 \end{vmatrix} = (5)(-2)(3)(-2) = 60$$

19. Addition of -2 times the first row to the second row yields

$$\begin{vmatrix} 1 & 1 & 1 \\ 2 & 2 & 2 \\ 3 & 3 & 3 \end{vmatrix} = \begin{vmatrix} 1 & 1 & 1 \\ 0 & 0 & 0 \\ 3 & 3 & 3 \end{vmatrix} = 0$$

20. Addition of the first row to the second row yields

$$\begin{vmatrix} 2 & 3 & 4 \\ -2 & -3 & 1 \\ 3 & 2 & 7 \end{vmatrix} = \begin{vmatrix} 2 & 3 & 4 \\ 0 & 0 & 5 \\ 3 & 2 & 7 \end{vmatrix} = (-5)\begin{vmatrix} 2 & 3 \\ 3 & 2 \end{vmatrix} = 25$$

21. Addition of -2 times the first row to the third row yields

$$\begin{vmatrix} 3 & -2 & 5 \\ 0 & 5 & 17 \\ 6 & -4 & 12 \end{vmatrix} = \begin{vmatrix} 3 & -2 & 5 \\ 0 & 5 & 17 \\ 0 & 0 & 2 \end{vmatrix} = (3)(5)(2) = 30$$

22. Addition of 3 times the second row to the first row yields

$$\begin{vmatrix} -3 & 6 & 5 \\ 1 & -2 & -4 \\ 2 & -5 & 12 \end{vmatrix} = \begin{vmatrix} 0 & 0 & -7 \\ 1 & -2 & -4 \\ 2 & -5 & 12 \end{vmatrix} = (-7)\begin{vmatrix} 1 & -2 \\ 2 & -5 \end{vmatrix} = 7$$

23. Addition of -2 times the first row to the 4th row yields

$$\begin{vmatrix} 1 & 2 & 3 & 4 \\ 0 & 5 & 6 & 7 \\ 0 & 0 & 8 & 9 \\ 2 & 4 & 6 & 9 \end{vmatrix} = \begin{vmatrix} 1 & 2 & 3 & 4 \\ 0 & 5 & 6 & 7 \\ 0 & 0 & 8 & 9 \\ 0 & 0 & 0 & 1 \end{vmatrix} = 40$$

24. Addition of twice the 1st row to the 4th row yields

$$\begin{vmatrix} 2 & 0 & 0 & -3 \\ 0 & 1 & 11 & 12 \\ 0 & 0 & 5 & 13 \\ -4 & 0 & 0 & 7 \end{vmatrix} = \begin{vmatrix} 2 & 0 & 0 & -3 \\ 0 & 1 & 11 & 12 \\ 0 & 0 & 5 & 13 \\ 0 & 0 & 0 & 1 \end{vmatrix} = 10$$

25. Multiplication of the second row by abc yields

$$\det A = \begin{vmatrix} 1 & 1 & 1 \\ bc & ac & ab \\ bc & ac & ab \end{vmatrix} = 0$$

26. Addition of the 2nd row to the 3rd row yields

$$\det A = \begin{vmatrix} 1 & 1 & 1 \\ a & a & a \\ a+b+c & a+b+c & a+b+c \end{vmatrix} = 0$$

27. Expand the lefthand determinant along its first column, and then factor k out of the result.

28. Expand each of the two 3×3 determinants along its 3rd row, and then apply the interchange property for 2×2 determinants.

29. Expand the lefthand determinant along its 3rd column.

30. Expand the lefthand determinant along its first column. Then use the fact that a determinant with two identical columns has value zero.

31. All three parts follow immediately from the definition of the transpose.

32. The ijth element of $(AB)^T$ is the jith element of AB, and hence is the product of the jth row of A and the ith column of B. The ijth element of B^TA^T is the product of the ith row of B^T and the jth column of A^T. Because transposing a matrix changes the ith row to the ith column and vice versa, it follows that the ijth element of B^TA^T is the product of the jth row of A and the ith column of B. Thus the matrices $(AB)^T$ and B^TA^T have the same ijth elements, and so are equal.

33. If $A = \begin{bmatrix} a & b & c \\ d & e & f \\ g & h & i \end{bmatrix}$, then $A^T = \begin{bmatrix} a & d & g \\ b & e & h \\ c & f & i \end{bmatrix}$.

Expansion of $\det A^T$ along the first row gives

$$\det A^T = a\begin{vmatrix} e & f \\ h & i \end{vmatrix} - b\begin{vmatrix} d & f \\ g & i \end{vmatrix} + c\begin{vmatrix} d & e \\ g & h \end{vmatrix}.$$

Expansion of $\det A^T$ along the first column gives

$$\det A^T = a\begin{vmatrix} e & h \\ f & i \end{vmatrix} - b\begin{vmatrix} d & g \\ f & i \end{vmatrix} + c\begin{vmatrix} d & g \\ e & h \end{vmatrix}$$

Hence it follows that $\det A = \det A^T$ by the analogous result for 2×2 determinants.

34. The formula in (9) follows directly upon expanding the determinant of $A = [a_{ij}]$ along its first row and then evaluating each of the three 2×2 determinants that result.

35. Subtracting the first row from both the second row and the third row, we get

$$\begin{vmatrix} 1 & a & a^2 \\ 1 & b & b^2 \\ 1 & c & c^2 \end{vmatrix} = \begin{vmatrix} 1 & a & a^2 \\ 0 & b-a & b^2-a^2 \\ 0 & c-a & c^2-a^2 \end{vmatrix}$$

$$= (b-a)(c^2-a^2) - (c-a)(b^2-a^2)$$

$$= (b-a)(c-a)[(c+a)-(b+a)]$$

$$= (b-a)(c-a)(c-b)$$

36. Expansion of the 4×4 determinant defining $P(y)$ along its 4th row yields

$$P(y) = y^3\begin{vmatrix} 1 & x_1 & x_1^2 \\ 1 & x_2 & x_2^2 \\ 1 & x_3 & x_3^2 \end{vmatrix} + \cdots$$

$$= y^3V(x_1,x_2,x_3) + \cdots.$$

Since it is clear from the determinant definition of $P(y)$ that $P(x_1) = P(x_2) = P(x_3) = 0$, the three roots of the cubic polynomial $P(y)$ are x_1, x_2, x_3. The factor theorem therefore says that

$$P(y) = k(y - x_1)(y - x_2)(y - x_3)$$

for some constant k, and the calculation above yields

$$k = V(x_1,x_2,x_3) = (x_3 - x_1)(x_3 - x_2)(x_2 - x_1).$$

Finally we see that

$$V(x_1,x_2,x_3,x_4) = P(x_4)$$

$$= V(x_1,x_2,x_3)(x_4 - x_1)(x_4 - x_2)(x_4 - x_3)$$

which is the desired formula for $V(x_1,x_2,x_3,x_4)$.

37. The same argument as in Problem 36 yields

$$P(y) = V(x_1,x_2,\cdots,x_{n-1})(y - x_1)(y - x_2)\cdots(y - x_{n-1}).$$

Therefore

$$V(x_1,x_2,\cdots,x_n) = P(x_n)$$

$$= (x_n - x_1)(x_n - x_2)\cdots(x_n - x_{n-1})V(x_1,x_2,\cdots,x_{n-1})$$

$$= (x_n - x_1)(x_n - x_2)\cdots(x_n - x_{n-1})\prod_{i>j}^{n-1} (x_i - x_j)$$

$$= \prod_{i>j}^{n} (x_i - x_j)$$

38. (a)

$(4 - 1)(4 - 2)(4 - 3)(3 - 1)(3 - 2)(2 - 1) = 12$

(b)

$[3 -(-1)][3 - 2][3 -(-2)][(-2) -(-1)][(-2)-2][2-(-1)] = 240$

SECTION 2.3

DETERMINANTS AND ELEMENTARY ROW OPERATIONS

This section applies the method of Gaussian elimination to reduce a square matrix to triangular form, and thereby to calculate its determinant much more rapidly than by cofactor expansion. Students should be advised to use cofactor expansions only in the case of 2 x 2 and 3 x 3 determinants, and to use the method of elimination for all larger determinants.

1.	10	2.	36
3.	-22	4.	29
5.	-6	6.	-3
7.	56	8.	-37
9.	78	10.	-22
11.	-74	12.	84
13.	8	14.	135
15.	39	16.	79
17.	-30	18.	-160
19.	114	20.	-98

21. $|B| = 3^2|A| = 18;\quad |B^{-1}| = 1/18;\quad |C| = 6^2|B^{-1}| = 2$

22. $|B| = 2^3|A| = 32;\quad |B^{-1}| = 1/32;\quad |C| = 4^3|B^{-1}| = 2$

23. $|A| + |B| = ad - bc + 1$

whereas

$|A + B| = (a + 1)(d + 1) - bc = ad - bc + 1 + (a + d)$

Thus it is clear that $|A + B| = |A| + |B|$ if and only if $a + d = 0$.

24. If $A^2 = A$ then $|A^2| = |A|$ so $|A|(|A| - 1) = 0$. Hence either $|A| = 0$ or $|A| = 1$.

25. $A^n = 0$ implies $|A|^n = 0$, so $|A| = 0$.

26. $A^T = A^{-1}$ implies $|A^T| = |A^{-1}| = 1/|A|$, so $|A|^2 = 1$, and hence $|A| = \pm 1$.

27. $|A| = |P^{-1}BP| = |P^{-1}||B||P| = |P|^{-1}|B||P| = |B|$

28. (a) If A is a 3×3 skew-symmetric matrix then

$$|A| = |A^T| = |-A| = (-1)^3|A| = -|A|$$

so it follows that $|A| = 0$.

(b) $A = \begin{bmatrix} 2 & 1 \\ -1 & 2 \end{bmatrix}$ is a skew-symmetric matrix with $|A| = 5 \neq 0$.

29. (a) $|E| = -1$ and EB is the result of interchanging two rows of B. Hence

$$|EB| = -|B| = (-1)|B| = |E||B|.$$

(b) $|E| = 1$ and EB is the result of adding one row of B to another. Hence

$$|EB| = |B| = (1)|B| = |E||B|.$$

30. If A and B are invertible, then $|A| \neq 0$ and $|B| \neq 0$. Hence $|AB| = |A||B| \neq 0$, so it follows that AB is invertible. Conversely, AB invertible implies $|A||B| \neq 0$, so both $|A| \neq 0$ and $|B| \neq 0$, and therefore A and B are both invertible.

31. If either $AB = I$ or $BA = I$ is given, then it follows from Problem 30 that A and B are both invertible because their product (one way or the other) is invertible. Hence A^{-1} exists. If (for instance) it is $AB = I$ that is given, then multiplication by A^{-1} on the left yields $B = A^{-1}$.

32. These are immediate computations. The point is to note that $3 = 2 + 1$ and $4 = 3 + 1$.

33. $$\begin{vmatrix} 2 & 1 & 0 & 0 \\ 1 & 2 & 1 & 0 \\ 0 & 1 & 2 & 1 \\ 0 & 0 & 1 & 2 \end{vmatrix} = 2\begin{vmatrix} 2 & 1 & 0 \\ 1 & 2 & 1 \\ 0 & 1 & 2 \end{vmatrix} - \begin{vmatrix} 1 & 1 & 0 \\ 0 & 2 & 1 \\ 0 & 1 & 2 \end{vmatrix}$$

$$= 2\begin{vmatrix} 2 & 1 & 0 \\ 1 & 2 & 1 \\ 0 & 1 & 2 \end{vmatrix} - \begin{vmatrix} 2 & 1 \\ 1 & 2 \end{vmatrix}$$

$$= 2(4) - (3) = 5$$

by the results in Problem 32.

34. (a) The recursion formula

$$B_n = 2B_{n-1} - B_{n-2}$$

results almost immediately when we expand the $n \times n$ determinant B_n in precisely the same way we expanded the 4×4 determinant of Problem 33.

(b) If we assume inductively that

$$B_{n-1} = (n - 1) + 1 = n \quad \text{and} \quad B_{n-2} = (n - 2) + 1 = n - 1,$$

then the recursion formula of part (a) yields

$$B_n = 2B_{n-1} - B_{n-2} = 2(n) - (n - 1) = n + 1.$$

SECTION 2.4

CRAMER'S RULE AND INVERSE MATRICES

Students should understand that Cramer's rule for the solution of $n \times n$ linear systems, and the adjoint method for finding A^{-1}, are not practical computational methods, and hence are largely of theoretical importance. Therefore the purpose of Problems 1-20 is mainly to reinforce the student's understanding and memory of the statements of Cramer's rule and the adjoint formula for A^{-1}.

In Problems 1-10 we give the determinant of the coefficient matrix A as well as the solution of the given linear system.

1. $|A| = 5$, $x_1 = -3/5$, $x_2 = -1$, $x_3 = 4/5$

2. $|A| = 9$, $x_1 = -10/9$, $x_2 = 28/9$, $x_3 = -17/9$

3. $|A| = 73$, $x_1 = 23/73$, $x_2 = -53/73$, $x_3 = -32/73$

4. $|A| = 35$, $x_1 = 4/7$, $x_2 = 3/7$, $x_3 = 2/7$

5. $|A| = 23$, $x_1 = 2$, $x_2 = 3$, $x_3 = 0$

6. $|A| = 56$, $x_1 = -1/7$, $x_2 = 9/14$, $x_3 = 2/7$

7. $|A| = 14$, $x_1 = -8/7$, $x_2 = -10/7$, $x_3 = 1/7$

8. $|A| = 6$, $x_1 = -7/3$, $x_2 = 9$, $x_3 = 8$

9. $|A| = 37$, $x_1 = 9/37$, $x_2 = -7/37$, $x_3 = -8/37$

10. $|A| = 23$, $x_1 = 8/23$, $x_2 = 5/23$, $x_3 = -1/23$

11. $\frac{1}{10}\begin{bmatrix} 2 & 4 & 2 \\ -5 & 0 & -10 \\ -6 & -2 & -6 \end{bmatrix}$

12. $\frac{1}{22}\begin{bmatrix} 6 & -4 & 1 \\ -6 & 4 & 10 \\ -8 & -2 & -5 \end{bmatrix}$

13. $\frac{1}{73}\begin{bmatrix} -11 & -2 & 4 \\ 19 & 30 & 13 \\ 28 & 25 & 23 \end{bmatrix}$

14. $\frac{1}{35}\begin{bmatrix} -2 & -3 & 12 \\ 9 & -4 & -19 \\ 13 & 2 & -8 \end{bmatrix}$

15. $\frac{1}{35}\begin{bmatrix} -15 & 25 & -26 \\ 10 & -5 & 8 \\ 15 & -25 & 19 \end{bmatrix}$

16. $\frac{1}{23}\begin{bmatrix} 5 & 20 & -17 \\ 10 & 17 & -11 \\ 1 & 4 & -8 \end{bmatrix}$

17. $\frac{1}{29}\begin{bmatrix} 11 & -14 & -15 \\ -17 & 19 & 10 \\ 18 & -15 & -14 \end{bmatrix}$

18. $\frac{1}{6}\begin{bmatrix} -6 & 10 & 2 \\ 15 & -21 & -6 \\ 12 & -18 & -6 \end{bmatrix}$

19. $\frac{1}{37}\begin{bmatrix} -21 & -1 & -13 \\ 4 & 9 & 6 \\ -6 & 5 & -9 \end{bmatrix}$

20. $\frac{1}{107}\begin{bmatrix} 9 & 12 & -13 \\ 11 & -21 & -4 \\ -15 & -20 & -14 \end{bmatrix}$

21. If $A = \begin{bmatrix} a & b & c \\ 0 & d & e \\ 0 & 0 & f \end{bmatrix}$ and $|A| = adf \neq 0$, then the cofactor matrix of A is $[\,A_{ij}\,] = \begin{bmatrix} df & 0 & 0 \\ -bf & af & 0 \\ be-cd & -ae & ad \end{bmatrix}$.

Therefore $A^{-1} = \frac{1}{adf}\begin{bmatrix} df & -bf & be-cd \\ 0 & af & -ae \\ 0 & 0 & ad \end{bmatrix}$.

22. (a) Follows immediately from Equation (19).
(b) Follows immediately from Equation (7).

23. If A is symmetric, then $A_{ji} = A_{ij}$, so the cofactor matrix of A is symmetric. Then adj $A = [A_{ij}\,]^T$ and hence $A^{-1} = (\text{adj } A)/|A|$ are also symmetric.

24. The coefficient determinant is

$$\begin{vmatrix} 0 & c & b \\ c & 0 & a \\ b & a & 0 \end{vmatrix} = -c\begin{vmatrix} c & a \\ b & 0 \end{vmatrix} + b\begin{vmatrix} c & 0 \\ b & a \end{vmatrix} = 2abc.$$

The numerator determinant for $\cos A$ is

$$\begin{vmatrix} a & c & b \\ b & 0 & a \\ c & a & 0 \end{vmatrix} = -b\begin{vmatrix} c & b \\ a & 0 \end{vmatrix} - a\begin{vmatrix} a & c \\ c & a \end{vmatrix}$$

$$= ab^2 - a^3 + ac^2.$$

Therefore $\cos A = \dfrac{ab^2 - a^3 + ac^2}{2abc} = \dfrac{b^2 + c^2 - a^2}{2bc}$.

SECTION 2.5

DETERMINANTS AND CURVE FITTING

Problems 1-3 are routine applications of Formula (2) in this section. Problems 4-6 are applications of Formula (4); Problems 7-10 are applications of Formula (6); and Problems 12-15 are applications of Formula (9).

1. $y = -2 + 3x$

2. $y = 4 - 7x$

3. $y = 3 - 2x^2$

4. $y = 2x + 3x^2$

5. $y = 5 - 3x + x^2$

6. $y = -10 - 7x + 2x^2$

7. $x^2 + y^2 - 6x - 4y = 12$

8. $x^2 + y^2 + 6x - 8y = 75$

9. $x^2 + y^2 + 4x + 4y = 5$

10. $x^2 + y^2 - 10x - 24y = 0$

11. There is an x^2 term, an xy term, a y^2 term, and a constant term. Moreover, the substitution of any one of the three given points yields a determinant with two identical rows.

12. $x^2 - xy + y^2 = 25$

13. $4x^2 - 7xy + 4y^2 = 100$

14. $100x^2 - 199xy + 100y^2 = 100$

15. $400x^2 - 481xy + 225y^2 = 3600$

16. In Figure 2.13 let E denote the point $(x_1,0)$ and F the point $(x_2,0)$. Then

$$\text{area } OAB = \text{area } OBF + \text{area } BFEA - \text{area } OAE$$

$$= \tfrac{1}{2}x_2y_2 + \tfrac{1}{2}(x_1 - x_2)(y_1 + y_2) - \tfrac{1}{2}x_1y_1$$

$$= \tfrac{1}{2}(x_1y_2 - x_2y_1)$$

$$\text{area } OAB = \frac{1}{2}\begin{vmatrix} x_1 & y_1 \\ x_2 & y_2 \end{vmatrix}$$

17. The parallelogram is the union of two congruent triangles, the area of each of which is given by the formula of Problem 16.

18. area ABC = area OAB - area OAC - area OBC

$$= \frac{1}{2}\begin{vmatrix} x_1 & y_1 \\ x_2 & y_2 \end{vmatrix} - \frac{1}{2}\begin{vmatrix} x_1 & y_1 \\ x_3 & y_3 \end{vmatrix} - \frac{1}{2}\begin{vmatrix} x_3 & y_3 \\ x_2 & y_2 \end{vmatrix}$$

$$\text{area } ABC = \frac{1}{2}\begin{vmatrix} x_1 & y_1 & 1 \\ x_2 & y_2 & 1 \\ x_3 & y_3 & 1 \end{vmatrix}$$

CHAPTER 3

VECTORS IN THE PLANE AND IN SPACE

This chapter discusses the concepts of linear independence and dependence and of orthogonality of vectors in the 2-dimensional plane and in 3-dimensional space. The purpose of this introductory discussion is to provide concrete background for the more general treatment of these concepts in Chapters 4 and 5. The latter two chapters are self-contained, however, so the student who is sufficiently familiar with vectors in R^2 and R^3 can proceed directly to Chapter 4 without loss of continuity.

SECTION 3.1

THE VECTOR SPACE R^2

The purpose of this section is to introduce the fundamental concepts of vectors, vector spaces, linear independence, and bases -- all in the relatively familiar and readily visualized context of the coordinate plane R^2.

Problems 1-8 ask for the values of $|\bar{a}|$, $|-2\bar{b}|$, $|\bar{a} - \bar{b}|$, $\bar{a} + \bar{b}$, and $3\bar{a} - 2\bar{b}$ for the given vectors $\bar{a}$ and $\bar{b}$. We list these values in tabular form:

	$\lvert\bar{a}\rvert$	$\lvert -2\bar{b}\rvert$	$\lvert\bar{a} - \bar{b}\rvert$	$\bar{a} + \bar{b}$	$3\bar{a} - 2\bar{b}$
1.	$\sqrt{5}$	$2\sqrt{13}$	$4\sqrt{2}$	(-2,0)	(9,-10)
2.	5	10	$5\sqrt{2}$	(-1,7)	(17,6)
3.	$2\sqrt{2}$	10	$\sqrt{5}$	(-5,-6)	(0,2)
4.	$2\sqrt{65}$	$12\sqrt{5}$	$20\sqrt{2}$	(4,-8)	(-48,-54)
5.	$\sqrt{10}$	$2\sqrt{29}$	$\sqrt{65}$	(3,-2)	(-1,19)
6.	$\sqrt{29}$	$2\sqrt{37}$	$\sqrt{2}$	(3,-11)	(4,-3)
7.	4	14	$\sqrt{65}$	(4,-7)	(12,14)

8. $\sqrt{2}$ $4\sqrt{2}$ $3\sqrt{2}$ $(1,1)$ $(-7,-7)$

In each of Problems 9-12, $\bar{v} = -\bar{u}$. We give only the value of the unit vector $\bar{u} = \bar{a}/|\bar{a}|$.

9. $|\bar{a}| = 5$, $\bar{u} = (-3/5,-4/5)$

10. $|\bar{a}| = 13$, $\bar{u} = (5/13,-12/13)$

11. $|\bar{a}| = 17$, $\bar{u} = (8/17,15/17)$

12. $|\bar{a}| = 25$, $\bar{u} = (7/25,-24/25)$

In each of Problems 13-16 we get the components of $\bar{a}$ by subtracting the coordinates of P from those of Q.

13. $(0,-4)$

14. $(0,1)$

15. $(8,-14)$

16. $(-5,0)$

17. $\bar{v} = (3/2)\bar{u}$, so the vectors are linearly dependent.

18. $a\bar{u} + b\bar{v} = (3b,2a) = (0,0)$ implies $a = b = 0$, so the vectors are linearly independent.

19. $a\bar{u} + b\bar{v} = (2a + 2b,\ 2a - b) = (0,0)$ implies $a = b = 0$, so the vectors are linearly independent.

20. $\bar{v} = -\bar{u}$, so the vectors are linearly dependent.

In each of Problems 21-26 we solve (as in Example 3) the system

$$\begin{bmatrix} u_1 & v_1 \\ u_2 & v_2 \end{bmatrix} \begin{bmatrix} a \\ b \end{bmatrix} = \begin{bmatrix} w_1 \\ w_2 \end{bmatrix}$$

for the coefficients a and b in $\bar{w} = a\bar{u} + b\bar{v}$.

21. $\bar{w} = 3\bar{u} + 2\bar{v}$

22. $\bar{w} = 2\bar{u} - 3\bar{v}$

23. $\bar{w} = \bar{u} - 2\bar{v}$

24. $\bar{w} = 3\bar{u} + 5\bar{v}$

25. $\bar{w} = 2\bar{u} - 3\bar{v}$

26. $\bar{w} = 7\bar{u} + 5\bar{v}$

27. (a)
$$\begin{aligned} \bar{u} + \bar{v} &= (u_1,u_2) + (v_1,v_2) \\ &= (u_1 + v_1,\ u_2 + v_2) \\ &= (v_1 + u_1,\ v_2 + u_2) \\ &= (v_1,v_2) + (u_1,u_2) \\ &= \bar{v} + \bar{u} \end{aligned}$$

(b)
$$\begin{aligned} \bar{u} + (\bar{v} + \bar{w}) &= (u_1,u_2) + [(v_1,v_2) + (w_1,w_2)] \\ &= (u_1 + (v_1 + w_1),\ u_2 + (v_2 + w_2)) \\ &= ((u_1 + v_1) + w_1,\ (u_2 + v_2) + w_2) \\ &= [(u_1,u_2) + (v_1,v_2)] + (w_1,w_2) \\ &= (\bar{u} + \bar{v}) + \bar{w} \end{aligned}$$

28. (a)
$$\begin{aligned} (r + s)\bar{u} &= (r + s)(u_1,u_2) \\ &= ((r + s)u_1,\ (r + s)u_2) \\ &= (ru_1 + su_1,\ ru_2 + su_2) \\ &= r(u_1,u_2) + s(u_1,u_2) \\ &= r\bar{u} + s\bar{u} \end{aligned}$$

(b)
$$\begin{aligned} (rs)\bar{u} &= r(s(u_1,u_2)) \\ &= r(su_1,su_2) \\ &= (rsu_1,rsu_2) \\ &= rs(u_1,u_2) = (rs)\bar{u} \end{aligned}$$

29. Let $A = \begin{bmatrix} u_1 & v_1 \\ u_2 & v_2 \end{bmatrix}$ and $\bar{x} = \begin{bmatrix} a \\ b \end{bmatrix}$.

We want to prove that the vectors $\bar{u} = (u_1, u_2)$ and $\bar{v} = (v_1, v_2)$ are linearly independent if and only if $\det A \neq 0$. First, $\bar{u}$ and $\bar{v}$ are linearly independent if and only if the equation

$$a\bar{u} + b\bar{v} = A\bar{x} = \bar{0}$$

implies that $a = b = 0$, that is, $\bar{x} = \bar{0}$. Thus $\bar{u}$ and $\bar{v}$ are linearly independent if and only if the equation $A\bar{x} = \bar{0}$ has only the trivial solution $\bar{x} = \bar{0}$. This is so if and only if A is invertible. But, by Theorem 2 in Section 2.3, A is invertible if and only if $\det A \neq 0$.

30. Given three vectors $\bar{u}, \bar{v}, \bar{w}$ in R^2, we want to show that there exist numbers a, b, c not all zero such that

$$a\bar{u} + b\bar{v} + c\bar{w} = \bar{0}.$$

There are two cases. If $\bar{u}$ and $\bar{v}$ are linearly dependent, then by definition there exist a and b not both zero such that

$$a\bar{u} + b\bar{v} = \bar{0}.$$

Then

$$a\bar{u} + b\bar{v} + 0\bar{w} = \bar{0},$$

so the vectors $\bar{u}, \bar{v}, \bar{w}$ are linearly dependent. If $\bar{u}$ and $\bar{v}$ are linearly independent, then by Theorem 3 there exist numbers a and b such that

$$\bar{w} = a\bar{u} + b\bar{v}.$$

Then

$$a\bar{u} + b\bar{v} + (-1)\bar{w} = \bar{0},$$

so again the vectors $\bar{u}, \bar{v}, \bar{w}$ are linearly dependent.

SECTION 3.2

THE VECTOR SPACE R^3

Here the fundamental vector space concepts introduced in Section 3.1 are illustrated in the case of the familiar 3-dimensional coordinate space R^3. In addition, the concept of a subspace of a vector space is discussed briefly, and elementary linear independence results are used to prove that the only (nontrivial) subspaces of R^3 are lines and planes through the origin.

Problems 1-4 ask for the values of $|\bar{a} - \bar{b}|$, $2\bar{a} + \bar{b}$, and $3\bar{a} - 4\bar{b}$ for the given vectors $\bar{a}$ and $\bar{b}$. We list these values in tabular form.

	$\|\bar{a} - \bar{b}\|$	$2\bar{a} + \bar{b}$	$3\bar{a} - 4\bar{b}$
1.	$\sqrt{51}$	(5,8,-11)	(2,23,0)
2.	9	(1,4,-1)	(-15,-16,26)
3.	$3\sqrt{21}$	(9,-3,3)	(-14,-21,43)
4.	$\sqrt{17}$	(4,-1,-3)	(6,-7,12)

In Problems 5-8 we calculate $\det[\bar{u}\ \bar{v}\ \bar{w}]$, the determinant of the 3×3 matrix having $\bar{u}, \bar{v}$, and $\bar{w}$ as its column vectors, in order to determine whether these vectors are linearly independent ($\det \neq 0$) or linearly dependent ($\det = 0$).

5. det = 0, linearly dependent

6. det = 0, linearly dependent

7. det = -5, linearly independent

8. det = 9, linearly independent

In each of Problems 9-14 we solve the system

$$\begin{bmatrix} u_1 & v_1 & w_1 \\ u_2 & v_2 & w_2 \\ u_3 & v_3 & w_3 \end{bmatrix}\begin{bmatrix} a \\ b \\ c \end{bmatrix} = \begin{bmatrix} 0 \\ 0 \\ 0 \end{bmatrix}$$

to find whether it has a nontrivial solution (a,b,c) such that $a\bar{u} + b\bar{v} + c\bar{w} = \bar{0}$. If not, the three vectors are linearly independent.

9. $3\bar{u} + 2\bar{v} + \bar{w} = \bar{0}$

10. The system

$$\begin{bmatrix} 5 & 2 & 4 \\ 5 & 3 & 1 \\ 4 & 1 & 5 \end{bmatrix}\begin{bmatrix} a \\ b \\ c \end{bmatrix} = \begin{bmatrix} 0 \\ 0 \\ 0 \end{bmatrix}$$

has the nontrivial solution $a = -2,\ b = 3,\ c = 1$ so $-2\bar{u} + 3\bar{v} + \bar{w} = \bar{0}$.

11. $11\bar{u} + 4\bar{v} - \bar{w} = \bar{0}$

12. The system

$$\begin{bmatrix} 1 & 5 & 0 \\ 1 & 1 & 1 \\ 0 & 3 & 2 \end{bmatrix}\begin{bmatrix} a \\ b \\ c \end{bmatrix} = \begin{bmatrix} 0 \\ 0 \\ 0 \end{bmatrix}$$

has only the trivial solution $a = b = c = 0$, so the vectors $\bar{u},\bar{v},\bar{w}$ are linearly independent.

13. linearly independent

14. linearly independent

In each of Problems 15-18 we solve the system

$$\begin{bmatrix} u_1 & v_1 & w_1 \\ u_2 & v_2 & w_2 \\ u_3 & v_3 & w_3 \end{bmatrix}\begin{bmatrix} a \\ b \\ c \end{bmatrix} = \begin{bmatrix} t_1 \\ t_2 \\ t_3 \end{bmatrix}$$

to find scalars a,b,c such that $\bar{t} = a\bar{u} + b\bar{v} + c\bar{w}$.

15. $\bar{t} = 2\bar{u} - \bar{v} + 3\bar{w}$

16. $\bar{t} = \bar{u} + 5\bar{v} - \bar{w}$

17. $\bar{t} = 2\bar{u} + 6\bar{v} + \bar{w}$

18. $\bar{t} = \bar{u} + \bar{v} + \bar{w}$

19. Given vectors $(0,y,z)$ and $(0,v,w)$ in V, we see that the sum $(0,y+v,z+w)$ and the scalar multiple $c(0,y,z) = (0,cy,cz)$ both have first component zero, and therefore are in V.

20. If (x,y,z) and $(u,v,w,)$ are in V, then

$$(x+u) + (y+v) + (z+w) = (x+y+z) + (u+v+w) = 0,$$

so the sum $(x+u,y+v,z+w)$ is in V. Similarly

$$cx + cy + cz = c(x + y + z) = 0,$$

so the scalar multiple (cx,cy,cz) is in V.

21. If (x,y,z) and (u,v,w) are in V, then

$$2(x + u) = 2x + 2u = 3y + 3v = 3(y + v),$$

so the sum $(x+u,y+v,z+w)$ is in V. Similarly

$$2(cx) = c(2x) = c(3y) = 3(cy),$$

so the scalar multiple (cx,cy,cz) is in V.

22. If (x,y,z) and (u,v,w) are in V, then

$$z + w = (2x + 3y) + (2u + 3v) = 2(x + u) + 3(y + v),$$

so the sum $(x+u,y+v,z+w)$ is in V. Similarly

$$cz = c(2x + 3y) = 2(cx) + 3(cy),$$

so the scalar multiple (cx,cy,cz) is in V.

23. $(0,1,0)$ is in V but the sum $(0,1,0) + (0,1,0) = (0,2,0)$ is not, so V is not closed under addition. Alternatively, $2(0,1,0) = (0,2,0)$ is not in V, so V is not closed under multiplication by scalars.

24. $(1,1,1)$ is in V, but

$$2(1,1,1) = (1,1,1) + (1,1,1) = (2,2,2)$$

is not, so V is closed neither under addition nor under multiplication by scalars.

25. V is closed under addition. However, $(0,0,1)$ is in V but $(-1)(0,0,1) = (0,0,-1)$ is not, so V is not closed under multiplication by scalars.

26. $(1,1,1)$ is in V, but

$$2(1,1,1) = (1,1,1) + (1,1,1) = (2,2,2)$$

is not, so V is closed neither under addition nor under multiplication by scalars.

27. Pick $\bar{u}$ in the (nonempty) vector space V. Then, with $c = 0$, the scalar multiple $c\bar{u} = 0\bar{u} = \bar{0}$ must be in V.

28. If $\bar{u}$ and $\bar{v}$ are in V, then so are the scalar multiples $a\bar{u}$ and $b\bar{v}$. But then the sum $a\bar{u} + b\bar{v}$ is also in V. Thus a subspace is closed under forming linear combinations. Conversely, suppose the subset V of R^3 is closed under forming linear combinations. With $a = b = 1$, we see that the sum $\bar{u} + \bar{v}$ of two vectors in V is also in V. With $a = c$ and $b = 0$, we see that the scalar multiple $c\bar{u}$ is in V. Thus V is a subspace of R^3. Consequently we see that a subset of R^3 is a subspace if and only if it is closed under forming linear combinations.

29. Since $\bar{u}$ and $\bar{v}$ are linearly dependent, we know that either $\bar{v} = c\bar{u}$ (in which case we are finished) or $\bar{u} = k\bar{v}$. In the latter case the fact that $\bar{u} \neq \bar{0}$ implies that $k \neq 0$, so we can divide by k to get $\bar{v} = (1/k)\bar{u} = c\bar{u}$.

30. It suffices to show that every vector $\bar{v}$ in V is a scalar multiple of the given nonzero vector $\bar{u}$. If $\bar{u}$ and $\bar{v}$ were linearly independent then (by Theorem 3 in Section 3.1) they would form a basis for R^2, and it would follow that $V = R^2$. Since V is a nontrivial subspace of R^2, the vectors $\bar{u}$ and $\bar{v}$ must therefore be linearly dependent. But $\bar{u} \neq \bar{0}$, so Problem 29 shows that $\bar{v} = c\bar{u}$ for some scalar c.

31. Since the vectors $\bar{u}, \bar{v}, \bar{w}$ are linearly dependent there exist scalars p, q, r not all zero such that

$$p\bar{u} + q\bar{v} + r\bar{w} = \bar{0}.$$

If $r = 0$, then p and q are scalars not both zero such that $p\bar{u} + q\bar{v} = \bar{0}$. But this contradicts the fact that $\bar{u}$ and $\bar{v}$ are linearly independent. Hence $r \neq 0$, so we can solve for

$$\bar{w} = (-p/r)\bar{u} + (-q/r)\bar{v} = a\bar{u} + b\bar{v}.$$

32. If the vectors $\bar{u}$ and $\bar{v}$ are in the intersection V of the subspaces V_1 and V_2, then $\bar{u} + \bar{v}$ is in V_1 because $\bar{u}$ and $\bar{v}$ are in V_1, and $\bar{u} + \bar{v}$ is in V_2 because $\bar{u}$ and $\bar{v}$ are in V_2. Hence $\bar{u} + \bar{v}$ is in V, and thus V is closed under addition. It follows similarly that V is closed under multiplication by scalars, and is therefore a subspace.

SECTION 3.3

ORTHOGONALITY AND THE DOT PRODUCT

In this section we emphasize the dot product, the linear independence of mutually arthogonal vectors, and the use of an appropriate orthogonal basis in finding orthogonal projections.

The cross product will not be used subsequently in the text, and therefore is discussed less extensively. The Cauchy-Schwarz and triangle inequalities are relegated to the problems here because they will be discussed in more detail in Chapter 5.

1. $\cos\theta = 4/(\sqrt{45})(\sqrt{14}), \quad \theta \approx 80.83^\circ$

2. $\cos\theta = -13/(\sqrt{5})(\sqrt{50}), \quad \theta \approx 145.30^\circ$

3. $\cos\theta = 0, \quad \theta = 90^\circ$

4. $\cos\theta = -16/(\sqrt{38})(\sqrt{83}), \quad \theta \approx 106.55^\circ$

5. $\cos A = 5/(\sqrt{17})(\sqrt{38}), \quad A \approx 78.65^\circ$
$\cos B = 12/(\sqrt{17})(\sqrt{45}), \quad B \approx 64.29^\circ$
$\cos C = 33/(\sqrt{38})(\sqrt{45}), \quad C \approx 37.06^\circ$

6. We take the unit cube in the first quadrant with opposite vertices $(0,0,0)$ and $(1,1,1)$. If θ is the angle between the diagonal with these vertices and the edge with vertices $(0,0,0)$ and $(1,0,0)$, then $\cos\theta = 1/\sqrt{3}$, so $\theta \approx 54.74^\circ$.

7. If θ is the angle between the edges joining $(0,0,0)$ to $(1,1,0)$ and $(1,0,1)$, then $\cos\theta = 1/(\sqrt{2})(\sqrt{2}) = 1/2$, so $\theta = 60^\circ$. Thus the faces of the regular tetrahedron are equilateral triangles.

8. If θ is the angle between the lines from the centroid $(1/2,1/2,1/2)$ to the vertices $(1,1,0)$ and $(1,0,1)$, then $\cos\theta = (-1/4)/(\sqrt{(3/4)})^2 = -1/3$, so $\theta \approx 109.47^\circ$.

9. $\bar{v} = (8/3)\bar{a} + (4/3)\bar{b} - \bar{c}$

10. $\bar{v} = (15/8)\bar{a} - (4/11)\bar{b} - (3/11)\bar{c}$

11. $\bar{v} = (1/3)\bar{a} + (5/11)\bar{b} - (1/33)\bar{c}$

12. $\bar{v} = (2/9)\bar{a} + (2/9)\bar{b} + (1/27)\bar{c}$

In Problems 13-16 we use the formulas

$$\bar{b}_{\parallel} = \frac{\bar{a}\cdot\bar{b}}{a\cdot a}\,\bar{a}, \qquad \bar{b}_{\perp} = \bar{b} - \bar{b}_{\parallel}$$

and write the answers in the form

$$\bar{b} = \bar{b}_{\parallel} + \bar{b}_{\perp}$$

13. $\bar{b} = (2,2,2) + (0,1,-1)$

14. $\bar{b} = (3,0,-3) + (1,3,1)$

15. $\bar{b} = (4,2,2) + (1,-1,-1)$

16. $\bar{b} = (0,2,6) + (5,6,-2)$

In Problems 17 and 18 we let $\bar{a} = \bar{u}$ and $\bar{b} = \bar{v}$. In Problems 19 and 20 we let $\bar{a} = \bar{u}$ and

$$\bar{b} = \bar{v} - \frac{\bar{u}\cdot\bar{v}}{u\cdot u}\,\bar{u}.$$

In either case we then calculate

$$\bar{p} = \frac{\bar{w}\cdot\bar{a}}{a\cdot a}\,\bar{a} + \frac{\bar{w}\cdot\bar{b}}{b\cdot b}\,\bar{b}.$$

Thus we are simply applying Equations (15) and (17).

17. $\bar{p} = (9/6)(1,2,1) + (0/3)(1,-1,1) = (3/2,3,3/2)$

18. $\bar{p} = (3/9)(2,2,-1) + (6/9)(2,-1,2) = (2,0,1)$

19. $\bar{p} = (12/3)(1,1,1) + (-1/2)(0,1,-1) = (4,7/2,9/2)$

20. $\bar{p} = (4/2)(1,0,-1) + (9/11)(1,3,1) = (31/11,27/11,-13/11)$

In each of Problems 21-24 we simply use the determinant in Equation (22) to calculate $\bar{c} = \bar{a} \times \bar{b}$.

21. $(5,-5,-5)$

22. $(-2,17,7)$

23. $(2,-4,2)$

24. $(4,-10,-31)$

25. If $\bar{u}$ and $\bar{v}$ are elements of W then $\bar{a}\cdot\bar{u} = \bar{a}\cdot\bar{v} = 0$. It follows that

$$\bar{a}\cdot(r\bar{u} + s\bar{v}) = r\bar{a}\cdot\bar{u} + s\bar{a}\cdot\bar{v} = 0,$$

so any linear combination of $\bar{u}$ and $\bar{v}$ is also in W. Therefore W is a subspace of R^3.

26. $|\bar{a}\cdot\bar{b}| = |\bar{a}||\bar{b}||\cos\theta| \leq |\bar{a}||\bar{b}|$ because $|\cos\theta| \leq 1$.

27. $$\begin{aligned} |\bar{a} + \bar{b}|^2 &= (\bar{a} + \bar{b})\cdot(\bar{a} + \bar{b}) \\ &= \bar{a}\cdot\bar{a} + \bar{b}\cdot\bar{b} + 2\bar{a}\cdot\bar{b} \\ &\leq |\bar{a}|^2 + |\bar{b}|^2 + 2|\bar{a}\cdot\bar{b}| \\ &\leq |\bar{a}|^2 + |\bar{b}|^2 + 2|\bar{a}||\bar{b}| \quad \text{(Cauchy-Schwarz)} \\ &= (|\bar{a}| + |\bar{b}|)^2 \end{aligned}$$

Thus

$$|\bar{a} + \bar{b}|^2 \leq (|\bar{a}| + |\bar{b}|)^2,$$

and we get the triangle inequality when we take square roots.

28. If $r\bar{a} + s\bar{b} + t\bar{c} = \bar{0}$ then we get $t\bar{c}\cdot\bar{c} = 0$ when we take the dot product of $\bar{c}$ with each side. Because $\bar{c} \neq \bar{0}$, it follows that $t = 0$, so the original equation reduces to $r\bar{a} + s\bar{b} = \bar{0}$. Since $\bar{a}$ and $\bar{b}$ are linearly independent, it follows that $r = s = 0$. Thus $r\bar{a} + s\bar{b} + t\bar{c} = \bar{0}$ implies that $r = s = t = 0$, so the vectors $\bar{a},\bar{b},\bar{c}$ are linearly independent.

29. Each vector $\bar{v}$ in V is a linear combination $\bar{v} = r\bar{a} + s\bar{b}$ of $\bar{a}$ and $\bar{b}$. Because $\bar{w}\cdot\bar{a} = \bar{w}\cdot\bar{b} = 0$, it follows that

$$\bar{w}\cdot\bar{v} = \bar{w}\cdot(r\bar{a} + s\bar{b}) = r\bar{w}\cdot\bar{a} + s\bar{w}\cdot\bar{b} = 0.$$

Thus $\bar{w}$ is orthogonal to every vector $\bar{v}$ in V, and is therefore orthogonal to V.

30. If Q_1 and Q_2 were two distinct points of V with the two segments PQ_1 and PQ_2 both orthogonal to V, then the triangle PQ_1Q_2 would have two right angles, which is impossible.

31. To the contrary, suppose that all three 2 x 2 submatrices have zero determinant. Then each of the following three pairs of 2-vectors is linearly dependent: (a_1,a_2) and (b_1,b_2), (a_2,a_3) and (b_2,b_3), and (a_1,a_3) and (b_1,b_3). There are several cases to consider, depending upon which vector in each pair is a multiple of the other. For instance, suppose that

$$(b_1,b_2) = r(a_1,a_2),\ (b_2,b_3) = s(a_2,a_3),\ (b_1,b_3) = t(a_1,a_3).$$

Then the fact that $b_2 = ra_2 = sa_2$ implies that either $r = s$ or $a_2 = 0$. Similarly, either $r = t$ or $a_1 = 0$, and either $s = t$ or $a_3 = 0$. But a_1,a_2,a_3 cannot all be zero (because $\bar{a}$ and $\bar{b}$ are linearly independent), so it follows that either $r = s$ or $r = t$ or $s = t$. If $r = s$, for instance, then $(b_1,b_2,b_3) = r(a_1,a_2,a_3)$, so $\bar{a}$ and $\bar{b}$ are linearly dependent, which is a contradiction.

32. $$\begin{aligned} |\bar{a} \times \bar{b}|^2 &= |\bar{a}|^2|\bar{b}|^2 - (\bar{a}\cdot\bar{b})^2 \\ &= |\bar{a}|^2|\bar{b}|^2 - |\bar{a}|^2|\bar{b}|^2 \cos^2\theta \\ &= |\bar{a}|^2|\bar{b}|^2(1 - \cos^2\theta) \end{aligned}$$

$$|\bar{a} \times \bar{b}|^2 = |\bar{a}|^2|\bar{b}|^2 \sin^2\theta$$

But $\sin\theta \geq 0$ because θ is between 0° and 180°. Hence we get $|\bar{a} \times \bar{b}| = |\bar{a}||\bar{b}|\sin\theta$ when we take square roots.

SECTION 3.4

LINES AND PLANES IN SPACE

Perhaps the greatest benefit of studying equations of lines and planes is the development of the ability to visualize various configurations of lines and planes in space. Students need to be reminded of the power of geometric visualization in algebra. For example, a picture of two planes makes it obvious that two linear equations have either no solution (if the planes are distinct and parallel) or infinitely many solutions (if they are coincident or nonparallel). In preparation for vector spaces and subspaces in Chapter 4, the fact that lines and planes are subspaces of R^3 should be emphasized.

1. $x = t,\ \ y = 2t,\ \ z = 3t$

2. $x = 3 - 2t,\ \ y = -4 + 7t,\ \ z = 5 + 3t$

3. $x = 4 + 2t,\ \ y = 13,\ \ z = -3 - 3t$

4. $x = -6t,\ \ y = 3t,\ \ y = 5t$

5. $x = 3 + 3t,\ \ y = 5 - 13t,\ z = 7 + 3t$

6. $x = t,\ \ y = t,\ \ z = t$

7. $x = 2 + 2t,\ \ y = -3 - t,\ \ z = 4 + 3t$

8. $x = 2 + 3t\ \ \ y = -1 + t,\ \ z = 5 - t$

9. $2x + 2y - z = 30$

10. $7x + 11y = 114$

11. $x - 3y + 2z = 14$

12. $y = 7$; the normal vector is $\bar{n} = (0,1,0)$.

13. $3x + 4y - z = 0$

14. $x + y - 2z = -2$

15. Because the plane passes through the origin, its equation is of the form $ax + by + cz = 0$. Substitution of the coordinates of P and of Q yields the equations

$$a + b + c = 0$$
$$a - b + 3c = 0$$

with general solution $b = c = t$, $a = -2t$. With $t = -1$ we get the equation

$$2x - y - z = 0$$

of the desired plane.

16. When we substitute in $ax + by + cz = d$ the coordinates of the three given points, we get the equations

$$\begin{aligned} a \qquad - c - d &= 0 \\ 3a + 3b + 2c - d &= 0 \\ 4a + 5b - c - d &= 0 \end{aligned}$$

with general solution $d = t$, $c = -t/16$, $b = -9t/16$, $a = 15t/16$. The choice $t = 16$ yields the equation of the desired plane

$$15x - 9y - z = 16.$$

17. The equations

$$5 + 3s = 3 - t, \quad -3 - 2s = -4 + 3t, \quad 4 + s = 5 - 2t$$

have the unique solution $s = -1$, $t = 1$, so the two lines intersect in the point $(2,-1,3)$.

18. The equations

$$4 + s = -4 + 3t, \quad -2 -2s = 4 - t, \quad 6 + 2s = -2 + 2t$$

have the unique solution $x = -2$, $t = 2$, so the two lines intersect in the point $(2,2,2)$.

19. The three equations in s and t are inconsistent, so the two lines do not intersect.

20. The three equations in s and t are inconsistent, so the two lines do not intersect.

In each of Problems 21-24 we find parametric equations of the line of intersection by solving the two given equations for x and y in terms of $z = t$.

21. $x = 2 - 2t$, $y = -1 - t$, $z = 3t$

22. $x = (7 - 2t)/5$, $y = (6 - t)/5$, $z = t$

23. $x = (10 - t)/2$, $y = -(10 - 5t)/2$, $z = t$

24. $x = 10$, $y = -10 - t$, $z = t$

Each of Problems 25-28 asks for the equation of a specified plane. In each of these problems we first find three points P_0, P_1, and P_2 that lie on the plane, and then use the method of Example 4 to find an equation of the plane. An alternative method would be to compute the normal vector

$$\bar{n} = \overrightarrow{P_0P_1} \times \overrightarrow{P_0P_2} \text{ (cross product)},$$

and then write the point-normal equation of the plane.

25. The plane $3x + 2y - 5z = 6$ intersects the plane $z = 0$ in the line $3x + 2y = 6$ in the xy-plane. Two points on this line are $(2,0)$ and $(0,3)$, so the desired plane is determined by the points $(1,1,1)$, $(2,0,0)$, and $(0,3,0)$. By the method of Example 4 we obtain the equation

$$3x + 2y + z = 6.$$

26. By inspection, we see that $(1,0,0)$ and $(1,1,1)$ lie on the line of intersection of the two given planes. Hence the desired plane is determined by the points $(1,3,-2)$, $(1,0,0)$, and $(1,1,1)$. By the method of Example 4 we obtain the equation $x = 1$ (if we are too obtuse to note in advance that all three points have x-coordinate 1).

27. By the method of Problems 17-20 we find that the two lines intersect when $s = 0$ and $t = -1$, and thus at the point $(1,-1,2)$. With $s = 1$ and $t = 0$ we find the additional point $(2,1,3)$ on the first line and the point $(2,2,4)$ on the second line. Finally, for the plane through these three points, the method of Example 4 yields the equation

$$x - y + z = 4.$$

28. By the method of Problems 21-24 we find that the line of intersection of the two planes is

$$x = (5 - 3t)/2, \quad y = -(1 - 5t), \quad z = t,$$

whose direction vector is $-1/4$ times that of the given line. The value $t = 1$ gives the point $(1,1,1)$ on the given line, while the values $t = 0$ and $t = 1$ give the points $(1,3,2)$ and $(7,-2,-2)$ on the given line. Finally, for the plane through these three points, the method of Example 4 yields the equation

$$x - 2y + 4z = 3.$$

29. Substitution of the parametric equations of the line into the equation of the plane yields the equation

$$3(3 + 2t) - 2(1 - 3t) + (5 + 4t) = 44$$

whose solution is $t = 2$. This value yields the point $(7,-5,13)$ of intersection.

30. The parametric equations of the orthogonal line are

$$x = x_0 + at, \quad y = y_0 + bt, \quad z = z_0 + ct.$$

Substitution in the equation of the plane yields the equation

$$a(x_0 + at) + b(y_0 + bt) + c(z_0 + ct) = d$$

with solution

$$t_1 = \frac{d - (ax_0 + by_0 + cz_0)}{a^2 + b^2 + c^2}.$$

The point of intersection has coordinates

$$x_1 = x_0 + at_1, \quad y_1 = y_0 + bt_1, \quad z_1 = z_0 + ct_1.$$

Then

$$D = [(x_1 - x_0)^2 + (y_1 - y_0)^2 + (z_1 - z_0)^2]^{1/2}$$

$$= (a^2t_1^2 + b^2t_1^2 + c^2t_1^2)^{1/2}$$

$$= |t_1|(a^2 + b^2 + c^2)^{1/2}$$

$$D = \frac{|ax_0 + by_0 + cz_0 - d|}{(a^2 + b^2 + c^2)^{1/2}}$$

31. (a) $D = 4$ (b) $D = 2$

32. If the point (x_0, y_0, z_0) lies on the first plane then $ax_0 + by_0 + cz_0 = d_1$. Hence the formula here follows immediately from the one in Problem 30.

33. The indicated cross product is

$$\bar{n} = \begin{vmatrix} \bar{i} & \bar{j} & \bar{k} \\ 1 & 3 & 2 \\ 2 & -1 & 3 \end{vmatrix} = 11\bar{i} + \bar{j} - 7\bar{k},$$

so the point-normal equation of the plane is

$$11(x - 2) + (y - 4) - 7(z + 3) = 0,$$

or

$$11x + y - 7z = 47.$$

34. If $\bar{w} = s\bar{u} + t\bar{v}$ then

$$\bar{w}\cdot\bar{n} = s\bar{u}\cdot\bar{n} + t\bar{v}\cdot\bar{n} = 0$$

because the cross product $\bar{n} = \bar{u} \times \bar{v}$ is orthogonal to both

$\bar{u}$ and $\bar{v}$, and thus $\bar{w}$ is orthogonal to $\bar{n}$. Conversely, suppose that $\bar{w}$ is orthogonal to $\bar{n} = \bar{u} \times \bar{v}$. By Problem 28 in Section 3.3, the vectors $\bar{n}, \bar{u}, \bar{v}$ are linearly independent and hence form a basis for R^3. Therefore there exist scalars r, s, t such that

$$\bar{w} = r\bar{n} + s\bar{u} + t\bar{v}.$$

When we take the dot product of each side with $\bar{n}$, we find that $r = 0$ (because $\bar{w}\cdot\bar{n} = 0$ and $\bar{n}\cdot\bar{n} \neq 0$). Hence $\bar{w} = s\bar{u} + t\bar{v}$, as desired.

CHAPTER 4

VECTOR SPACES

The treatment of vector spaces in this chapter is very concrete. Prior to the final optional section of the chapter, essentially the only vector spaces that appear in the examples and problems are subspaces of Euclidean spaces, that is, vector spaces of n-tuples of real numbers. Our motivation throughout is the fact that the solution space of a homogeneous linear system $A\bar{x} = \bar{0}$ is precisely such a "concrete" vector space.

SECTION 4.1

THE VECTOR SPACE R^n AND SUBSPACES

We begin with the vector space R^n of n-tuples of real numbers. Taking advantage of our detailed discussion in Chapter 3 of the vector spaces R^2 and R^3, we move on quickly to the concept of a <u>subspace</u> of a vector space. The main objective in this section is for the student to understand what types of subsets of R^n are subspaces, and that our first reason for studying subspaces is the fact that the <u>solution space</u> of a homogeneous linear system $A\bar{x} = \bar{0}$ is a subspace of R^n (assuming that A is an $m \times n$ matrix).

1. If $\bar{u} = (u_1,u_2,0)$ and $\bar{v} = (v_1,v_2,0)$ are vectors in W, then the vectors

$$\bar{u} + \bar{v} = (u_1 + v_1,\ u_2 + v_2,\ 0)$$

and $c\bar{u} = (cu_1,cu_2,0)$ are also in W. Hence W is a subspace of R^3.

2. If $\bar{u} = (5u_2,u_2,u_3)$ and $\bar{v} = (5v_2,v_2,v_3)$ are vectors in W, then the vectors

$$\bar{u} + \bar{v} = (5u_2 + 5v_2,\ u_2 + v_2,\ u_3 + v_3)$$

and $c\bar{u} = (5cu_2, cu_2, cu_3)$ are also in W. Hence W is a subspace of R^3.

3. The vector $\bar{u} = (1,1,1)$ is in W, but the scalar multiple $2\bar{u} = (2,2,2)$ is not. Hence W is not a subspace of R^3.

4. The vectors $\bar{u} = (1,0,0)$ and $\bar{v} = (0,1,0)$ are in W, but their sum $\bar{u} + \bar{v} = (1,1,0)$ is not. Hence W is not a subspace of R^2.

5. If $\bar{u} = (u_1, u_2, u_3, u_4)$ and $\bar{v} = (v_1, v_2, v_3, v_4)$ are in W, then

$$(u_1 + v_1) + 2(u_2 + v_2) + 3(u_3 + v_3) + 4(u_4 + v_4)$$

$$= (u_1 + 2u_2 + 3u_3 + 4u_4) + (v_1 + 2v_2 + 3v_3 + 4v_4)$$

$$= 0 + 0 = 0.$$

Thus the sum $\bar{u} + \bar{v}$ is in W, and similarly $c\bar{u}$ is in W. Hence W is a subspace of R^4.

6. If $\bar{u} = (3u_3, 4u_4, u_3, u_4)$ and $\bar{v} = (3v_3, 4v_4, v_3, v_4)$ are vectors in W, then the vectors

$$\bar{u} + \bar{v} = (3u_3 + 3v_3,\ 4u_4 + 4v_4,\ u_3 + v_3,\ u_4 + v_4)$$

and $c\bar{u} = (3cu_3, 4cu_4, cu_3, cu_4)$ are also in W. Hence W is a subspace of R^4.

7. The vectors $\bar{u} = (1,1)$ and $\bar{v} = (1,-1)$ are in W, but their sum $\bar{u} + \bar{v} = (2,0)$ is not. Hence W is not a subspace of R^2.

8. W is simply the zero subspace of R^2.

9. The vector $\bar{u} = (1,0)$ is in W, but the scalar multiple $2\bar{u} = (2,0)$ is not. Hence W is not a subspace of R^2.

10. The vectors $\bar{u} = (1,0)$ and $\bar{v} = (0,1)$ are in W, but their

sum $\bar{u} + \bar{v} = (1,1)$ is not. Hence W is not a subspace of R^2.

11. If $\bar{u} = (u_1,u_2,u_3,u_4)$ and $\bar{v} = (v_1,v_2,v_3,v_4)$ are vectors in W, let

$$\bar{w} = \bar{u} + \bar{v} = (u_1 + v_1, u_2 + v_2, u_3 + v_3, u_4 + v_4).$$

Then

$$\begin{aligned} w_1 + w_2 &= (u_1 + v_1) + (u_2 + v_2) \\ &= (u_1 + u_2) + (v_1 + v_2) \\ &= (u_3 + u_4) + (v_3 + v_4) \\ &= (u_3 + v_3) + (u_4 + v_4) = w_3 + w_4. \end{aligned}$$

Thus $\bar{w} = \bar{u} + \bar{v}$ is in W, and similarly $c\bar{u}$ is in W. Hence W is a subspace of R^4.

12. The vectors $\bar{u} = (1,0,1,0)$ and $\bar{v} = (0,2,0,3)$ are in W, but their sum $\bar{u} + \bar{v} = (1,2,1,3)$ is not. Hence W is not a subspace of R^4.

13. The vectors $\bar{u} = (0,1,1,1)$ and $\bar{v} = (1,1,1,0)$ are in W, but their sum $\bar{u} + \bar{v} = (1,2,2,1)$ is not. Hence W is not a subspace of R^4.

14. The vector $\bar{u} = (1,1,1,1)$ is in W, but the scalar multiple $0\bar{u} = (0,0,0,0)$ is not. Hence W is not a subspace of R^4.

15. The reduced echelon form of the coefficient matrix is

$$\begin{bmatrix} 1 & 0 & 1 & 4 \\ 0 & 1 & 0 & 2 \\ 0 & 0 & 0 & 0 \end{bmatrix}.$$

The free variables are $x_3 = s$ and $x_4 = t$, and then $x_2 = -2t$, $x_1 = -s - 4t$. Thus the solution is

$$\begin{aligned} \bar{x} &= (-s - 4t,-2t,s,t) \\ &= s(-1,0,1,0) + t(-4,-2,0,1). \end{aligned}$$

16. The reduced echelon form of the coefficient matrix is

$$\begin{bmatrix} 1 & 0 & 1 & 5 \\ 0 & 1 & 1 & 3 \\ 0 & 0 & 0 & 0 \end{bmatrix}.$$

With $x_3 = s$ and $x_4 = t$ we get $x_2 = -s - 3t$ and $x_1 = -s - 5t$. Thus the solution is

$$\begin{aligned} \bar{x} &= (-s - 5t, -s - 3t, s, t) \\ &= s(-1,-1,1,0) + t(-5,-3,0,1). \end{aligned}$$

17. The reduced echelon form of the coefficient matrix is

$$\begin{bmatrix} 1 & 0 & -1 & 2 \\ 0 & 1 & 3 & -1 \\ 0 & 0 & 0 & 0 \end{bmatrix}.$$

With $x_3 = s$ and $x_4 = t$ we get $x_2 = -3s + t$ and $x_1 = s - 2t$. Thus

$$\begin{aligned} \bar{x} &= (s - 2t, -3s + t, s, t) \\ &= s(1,-3,1,0) + t(-2,1,0,1). \end{aligned}$$

18. The reduced echelon form of the coefficient matrix is

$$\begin{bmatrix} 1 & 0 & 0 & -2 & -3 \\ 0 & 1 & 0 & 1 & 4 \\ 0 & 0 & 1 & 2 & -5 \end{bmatrix}.$$

With $x_4 = s$ and $x_5 = t$ we get $x_3 = -2s + 5t$, $x_2 = -s - 4t$, and $x_1 = 2s + 3t$. Hence

$$\begin{aligned} \bar{x} &= (2s + 3t, -s - 4t, -2s + 5t, s, t) \\ &= s(2,-1,-2,1,0) + t(3,-4,5,0,1). \end{aligned}$$

19. The reduced echelon form of the coefficient matrix is

$$\begin{bmatrix} 1 & 0 & 1 & 0 \\ 0 & 1 & 2 & 0 \\ 0 & 0 & 0 & 1 \end{bmatrix}.$$

Thus $x_4 = 0$, and with $x_3 = t$ we get $x_2 = -2t$ and $x_1 = -t$. Hence the solution is

$$\bar{x} = (-t,-2t,t,0) = t(-1,-2,1,0).$$

20. The reduced echelon form of the coefficient matrix is

$$\begin{bmatrix} 1 & 0 & 0 & 5 \\ 0 & 1 & 0 & -3 \\ 0 & 0 & 1 & 2 \end{bmatrix}.$$

With $x_4 = t$ we get $x_3 = -2t$, $x_2 = 3t$ and $x_1 = -5t$. Hence the solution is

$$\bar{x} = (-5t,3t,-2t,t) = t(-5,3,-2,1)$$

21. The reduced echelon form of the coefficient matrix is

$$\begin{bmatrix} 1 & 0 & 0 & 3 \\ 0 & 1 & 0 & -2 \\ 0 & 0 & 1 & 4 \end{bmatrix}.$$

With $x_4 = t$ we get $x_3 = -4t$, $x_2 = 2t$, and $x_1 = -3t$. Hence the solution is

$$\bar{x} = (-3t,2t,-4t,t) = t(-3,2,-4,1).$$

22. The reduced echelon form of the coefficient matrix is

$$\begin{bmatrix} 1 & 0 & 0 & 6 \\ 0 & 1 & 0 & -4 \\ 0 & 0 & 1 & 3 \end{bmatrix}.$$

With $x_4 = t$ we get $x_3 = -3t$, $x_2 = 4t$, and $x_1 = -6t$. Hence the solution is

$$\bar{x} = (-6t,4t,-3t,t) = t(-6,4,-3,1)$$

23. The subspace W is nonempty and hence contains a vector $\bar{u}$. Therefore it contains the vector

$$(1)\bar{u} + (-1)\bar{u} = \bar{u} + (-\bar{u}) = \bar{0},$$

using Problem 24(c) below.

24. (a) The fact that

$$\bar{u} = 1\bar{u} = (1 + 0)\bar{u} = 1\bar{u} + 0\bar{u} = \bar{u} + 0\bar{u}$$

implies that $0\bar{u} = \bar{0}$, adding $-\bar{u}$ to both sides.

(b) The fact that

$$c\bar{0} = c(\bar{0} + \bar{0}) = c\bar{0} + c\bar{0}$$

implies that $c\bar{0} = \bar{0}$, adding $-c\bar{0}$ to both sides.

(c) The fact that

$$\bar{u} + (-1)\bar{u} = [1 + (-1)]\bar{u} = 0\bar{u} = \bar{0}$$

means that $(-1)\bar{u} = -\bar{u}$.

25. If W is a subspace, then it contains the scalar multiples $a\bar{u}$ and $b\bar{v}$, and hence contains their sum $a\bar{u} + b\bar{v}$. Conversely, if the subset W is closed under taking linear combinations of pairs of vectors, then it contains

$$(1)\bar{u} + (1)\bar{v} = \bar{u} + \bar{v} \quad \text{and} \quad (c)\bar{u} + (0)\bar{v} = c\bar{u},$$

and hence is a subspace.

26. The sum of two scalar multiples of $\bar{u}$ is a scalar multiple of $\bar{u}$, as is any scalar multiple of a scalar multiple of $\bar{u}$.

27. Let $a_1\bar{u} + b_1\bar{v}$ and $a_2\bar{u} + b_2\bar{v}$ be two vectors in $W = \{a\bar{u} + b\bar{v}\}$. Then

$$(a_1\bar{u} + b_1\bar{v}) + (a_2\bar{u} + b_2\bar{v}) = (a_1 + a_2)\bar{u} + (b_1 + b_2)\bar{v}$$

and

$$c(a_1\bar{u} + b_1\bar{v}) = (ca_1)\bar{u} + (ca_2)\bar{v}$$

are also in W. Thus W is a subspace.

28. If $\bar{u}$ and $\bar{v}$ are vectors in W, then $A\bar{u} = k\bar{u}$ and $A\bar{v} = k\bar{v}$ so

$$A(\bar{u} + \bar{v}) = A\bar{u} + A\bar{v} = k\bar{u} + k\bar{v} = k(\bar{u} + \bar{v}).$$

Thus $\bar{u} + \bar{v}$ is in W also. Similarly, $c\bar{u}$ is in W, so W is a subspace of R^n.

29. If $A\bar{x} = \bar{b}$ then

$$A\bar{y} = A(\bar{x} - \bar{x}_0) = A\bar{x} - A\bar{x}_0 = \bar{b} - \bar{b} = \bar{0}.$$

Conversely, $A(\bar{x} - \bar{x}_0) = \bar{0}$ implies that

$$A\bar{x} = A\bar{x}_0 = \bar{b}.$$

30. Let W denote the intersection of the subspaces U and V. If $\bar{u}$ and $\bar{v}$ are two vectors in W, then $\bar{u}$ and $\bar{v}$ are both in U so $\bar{u} + \bar{v}$ is in U, and also $\bar{u}$ and $\bar{v}$ are both in V so $\bar{u} + \bar{v}$ is in V. Therefore $\bar{u} + \bar{v}$ is in W, as is $c\bar{u}$ (similarly). Hence Theorem 1 implies that W is a subspace. If U and V are non-coincident planes through the origin in R^3, then their intersection W is a line through the origin.

31. Let $\bar{w}_1$ and $\bar{w}_2$ be two vectors in $U + V$. Then

$$\bar{w}_i = \bar{u}_i + \bar{v}_i \qquad (i = 1,2)$$

where $\bar{u}_1$ and $\bar{u}_2$ are in U, and $\bar{v}_1$ and $\bar{v}_2$ are in V. Then

$$\bar{w}_1 + \bar{w}_2 = (\bar{u}_1 + \bar{v}_1) + (\bar{u}_2 + \bar{v}_2)$$

$$= (\bar{u}_1 + \bar{u}_2) + (\bar{v}_1 + \bar{v}_2)$$

is in $U + V$ also. Similarly $c\bar{w}_1$ is in $U + V$, so it follows that $U + V$ is a subspace. If U and V are non-coincident lines through the origin in R^3, then $U + V$ is a plane through the origin.

SECTION 4.2

LINEAR COMBINATIONS AND INDEPENDENCE

In this section we use two types of computational problems as aids in understanding linear independence and dependence. The first problem is that of expressing a vector $\bar{w}$ as a linear combination of k given vectors (if possible), and the second is that of determining whether or not k given vectors are linearly independent. For vectors in R^n, each of these problems reduces to solving a linear system of n equations in k unknowns. This computational approach renders the concepts of linear independence and dependence more concrete and less abstract.

1. Linearly dependent, because $\bar{v}_2 = (3/2)\bar{v}_1$

2. Linearly independent, because obviously the two vectors are not scalar multiples of each other.

3. Any 3 vectors in R^2 are linearly dependent.

4. Any 4 vectors in R^3 are linearly dependent.

5. The equation

$$c_1\bar{v}_1 + c_2\bar{v}_2 + c_3\bar{v}_3 = (c_1, -2c_2, 3c_3) = (0,0,0)$$

implies that $c_1 = c_2 = c_3 = 0$, so the three vectors are linearly independent.

6. Obviously the determinant of the 3×3 matrix $A = [\bar{v}_1 \ \ \bar{v}_2 \ \ \bar{v}_3]$ is 1, not zero, so by Theorem 2 the three vectors are linearly independent.

7. The equation

$$c_1\bar{v}_1 + c_2\bar{v}_2 + c_3\bar{v}_3 = (2c_1 + 3c_2 + 4c_3, c_1, c_2, c_3) = \bar{0}$$

immediately yields $c_1 = c_2 = c_3 = 0$, so the three vectors are linearly independent.

8. Obviously $\bar{v}_3 = \bar{v}_1 + \bar{v}_2$, so the three vectors are linearly dependent.

9. The vector equation $c_1\bar{v}_1 + c_2\bar{v}_2 = \bar{w}$ yields the linear system

$$\begin{aligned} 5c_1 + 3c_2 &= 1 \\ 3c_1 + 2c_2 &= 0 \\ 4c_1 + 5c_2 &= -7 \end{aligned}$$

with solution $c_1 = 2$, $c_2 = -3$. Hence $\bar{w} = 2\bar{v}_1 - 3\bar{v}_2$.

10. $\bar{w} = 7\bar{v}_1 + 4\bar{v}_2$

11. $\bar{w} = \bar{v}_1 - 2\bar{v}_2$

12. $\bar{w} = 2\bar{v}_1 + 5\bar{v}_2$

13. The linear system

$$\begin{aligned} c_1 + 5c_2 &= 5 \\ 5c_1 - 3c_2 &= 2 \\ -3c_1 + 4c_2 &= -2 \end{aligned}$$

is inconsistent, so $\bar{w}$ is not a linear combination of $\bar{v}_1$ and $\bar{v}_2$.

14. $\bar{w}$ is not a linear combination of $\bar{v}_1$ and $\bar{v}_2$.

15. The vector equation $c_1\bar{v}_1 + c_2\bar{v}_2 + c_3\bar{v}_3 = \bar{w}$ yields the linear system

$$\begin{aligned} 2c_1 + 3c_2 + c_3 &= 4 \\ -c_1 \qquad + 2c_3 &= 5 \\ 4c_1 + c_2 - c_3 &= 6 \end{aligned}$$

with solution $c_1 = 3$, $c_2 = -2$, $c_3 = 4$. Hence $\bar{w} = 3\bar{v}_1 - 2\bar{v}_2 + 4\bar{v}_3$.

16. $\bar{w} = 6\bar{v}_1 - 2\bar{v}_2 + 3\bar{v}_3$

17. The three vectors are linearly independent.

18. The vector equation $c_1\bar{v}_1 + c_2\bar{v}_2 + c_3\bar{v}_3 = \bar{0}$ yields the linear system

$$\begin{aligned} 2c_1 + 4c_2 - 2c_3 &= 0 \\ -5c_2 + c_3 &= 0 \\ -3c_1 - 6c_2 + 3c_3 &= 0 \end{aligned}$$

with solution $c_1 = 3t/5$, $c_2 = t/5$, $c_3 = t$. With $t = 5$ we get $c_1 = 3$, $c_2 = 1$, $c_3 = 5$, so $3\bar{v}_1 + \bar{v}_2 + 5\bar{v}_3 = \bar{0}$.

19. The homogeneous linear system

$$\begin{aligned} 2c_1 + 5c_2 + 2c_3 &= 0 \\ 4c_2 - c_3 &= 0 \\ 3c_1 - 2c_2 + c_3 &= 0 \\ c_2 - c_3 &= 0 \end{aligned}$$

has only the trivial solution $c_1 = c_2 = c_3 = 0$, so the 3 vectors are linearly independent.

20. The three vectors are linearly independent.

21. $\bar{v}_1 - 2\bar{v}_2 - \bar{v}_3 = \bar{0}$

22. $7\bar{v}_1 + 5\bar{v}_2 - 9\bar{v}_3 = \bar{0}$

23. Because $\bar{v}_1$ and $\bar{v}_2$ are linearly independent, the vector equation

$$c_1\bar{u}_1 + c_2\bar{u}_2 = (c_1 + c_2)\bar{v}_1 + (c_1 - c_2)\bar{v}_2 = \bar{0}$$

yields the linear system

$$\begin{aligned} c_1 + c_2 &= 0 \\ c_1 - c_2 &= 0 \end{aligned}$$

having only the trivial solution $c_1 = c_2 = 0$.

24. The linear system

$$\begin{aligned} c_1 + 2c_2 &= 0 \\ c_1 + 3c_2 &= 0 \end{aligned}$$

has only the trivial solution $c_1 = c_2 = 0$.

25. Because $\bar{v}_1, \bar{v}_2, \bar{v}_3$ are linearly independent, the vector equation

$$c_1\bar{u}_1 + c_2\bar{u}_2 + c_3\bar{v}_3$$

$$= (c_1 + c_2 + c_3)\bar{v}_1 + (2c_2 + 2c_2)\bar{v}_2 + 3c_3\bar{v}_3 = \bar{0}$$

yields the linear system

$$\begin{aligned} c_1 + c_2 + c_3 &= 0 \\ 2c_2 + 2c_3 &= 0 \\ 3c_3 &= 0 \end{aligned}$$

having only the trivial solution $c_1 = c_2 = c_3 = 0$.

26. The linear system

$$\begin{aligned} c_2 + c_3 &= 0 \\ c_1 \quad + c_3 &= 0 \\ c_1 + c_2 \quad &= 0 \end{aligned}$$

has only the trivial solution $c_1 = c_2 = c_3 = 0$.

27. If the elements of S are $\bar{v}_1, \bar{v}_2, \cdots, \bar{v}_k$ with $\bar{v}_1 = \bar{0}$, then we can take $c_1 = 1$ and $c_2 = \cdots = c_k = 0$. Then $c_1, c_2, \cdots, c_k$ are coefficients not all zero such that

$$c_1\bar{v}_1 + c_2\bar{v}_2 + \cdots + c_k\bar{v}_k = \bar{0}.$$

28. Because the set S is linearly dependent, there exist scalars $c_1, c_2, \cdots, c_k$ not all zero so that

$$c_1\bar{v}_1 + c_2\bar{v}_2 + \cdots + c_k\bar{v}_k = \bar{0}.$$

If $c_{k+1} = \cdots = c_m = 0$ then

$$c_1\bar{v}_1 + c_2\bar{v}_2 + \cdots + c_m\bar{v}_m = \bar{0}$$

with the coefficients not all zero.

29. If some subset of S were linearly dependent, then Problem 28 would imply that S itself is linearly dependent.

30. Because U is a subspace and contains the vectors $\bar{v}_1, \bar{v}_2, \cdots, \bar{v}_k$, it contains every linear combination of them. But W consists solely of linear combinations of these vectors, so U contains W.

31. If S is contained in span(T), then every vector in S is a linear combination of vectors in T. Hence every vector in span(S) is a linear combination of linear combinations of vectors in T -- hence is a linear combination of vectors in T -- and therefore is in span(T).

32. If $\bar{u}$ is another vector in S then the $k+1$ vectors $\bar{v}_1, \bar{v}_2, \cdots, \bar{v}_k, \bar{u}$ are linearly dependent. Hence there exist scalars $c_1, c_2, \cdots, c_k, c$ not all zero such that

$$c_1\bar{v}_1 + c_2\bar{v}_2 + \cdots + c_k\bar{v}_k + c\bar{u} = \bar{0}.$$

If $c = 0$ then we have a contradiction to the hypothesis that $\bar{v}_1, \bar{v}_2, \cdots, \bar{v}_k$ are linearly independent. Hence $c \neq 0$, so we can solve for $\bar{u}$ as a linear combination of $\bar{v}_1, \bar{v}_2, \cdots, \bar{v}_k$.

33. The determinant of the $k \times k$ identity submatrix is nonzero, so this follows immediately from Theorem 3.

34. $|AB| = |A||B| \neq 0$, so Theorem 2 implies that the column vectors of AB are linearly independent.

35. Because the vectors $\bar{v}_1, \bar{v}_2, \cdots, \bar{v}_k$ are linearly independent,

6. Theorem 3 implies that some $k \times k$ submatrix A_0 of A has nonzero determinant. Let A_0 consist of the rows $i_1, i_2, \cdots, i_k$ of A, and let C_0 be the $k \times k$ submatrix consisting of the same rows of the product matrix AB. Then $C_0 = A_0B$, so $|C_0| = |A_0||B| \neq 0$, and hence Theorem 3 implies that the column vectors of AB are linearly independent.

SECTION 4.3

BASES FOR VECTOR SPACES

The concepts of basis and dimension are central to the understanding of finite-dimensional vector spaces. In particular, a basis $\{\bar{v}_1, \bar{v}_2, \cdots, \bar{v}_k\}$ for a subspace W of R^n enables us to visualize W as a k-dimensional "plane" through the origin in R^n. This is especially useful when W is the solution space of a homogeneous linear system, in which case a basis is a maximal linearly independent set of solutions of the system, and every other solution is a linear combination of these particular solutions.

1. A basis for R^2, because $\bar{v}_1$ and $\bar{v}_2$ are obviously linearly independent.

2. Not a basis for R^3, because $\bar{v}_2 = 2\bar{v}_1$.

3. Not a basis for R^3, because any four vectors in R^3 are linearly dependent.

4. Not a basis for R^4, because every basis for R^4 contains four vectors.

5. Not a basis for R^3, because any linear combination $\bar{x}$ of $\bar{v}_1, \bar{v}_2, \bar{v}_3$ has $x_1 = 0$.

6. $\det[\bar{v}_1 \;\; \bar{v}_2 \;\; \bar{v}_3] = -1 \neq 0$, so the three vectors are linearly independent, and hence form a basis for R^3.

7. $\det[\bar{v}_1 \;\; \bar{v}_2 \;\; \bar{v}_3] = 1 \neq 0$, so the three vectors form a basis

for R^3.

8. $\det[\bar{v}_1 \;\; \bar{v}_2 \;\; \bar{v}_3 \;\; \bar{v}_4] = 66 \neq 0$, so the four vectors form a basis for R^4.

9. The single equation $x - 2y + 5z = 0$ is already a system in reduced echelon form, with free variables y and z. With $y = s$, $z = t$, $x = 2s - 5t$ we get the basis vectors $\bar{v}_1 = (2,1,0)$ and $\bar{v}_2 = (-5,0,1)$.

10. With $x = s$ and $y = z = t$ we get the basis vectors $\bar{v}_1 = (1,0,0)$ and $\bar{v}_2 = (0,1,1)$.

11. The system of two equations has coefficient matrix

$$\begin{bmatrix} 1 & -2 & 5 \\ 0 & 1 & -1 \end{bmatrix}$$

already in echelon form. With $y = z = t$ and $x = -3t$ we get the single basis vector $\bar{v}_1 = (-3,1,1)$.

12. With $b = r$, $c = s$, $d = t$ and $a = r + s + t$ we get the basis vectors $\bar{v}_1 = (1,1,0,0)$, $\bar{v}_2 = (1,0,1,0)$, and $\bar{v}_3 = (1,0,0,1)$.

13. With $c = s$, $d = t$, $a = 3s$, $b = 4t$ we get the basis vectors $\bar{v}_1 = (3,0,1,0)$ and $\bar{v}_2 = (0,4,0,1)$.

14. With $b = s$, $d = t$, $a = -2s$, $c = -3t$ we get the basis vectors $\bar{v}_1 = (-2,1,0,0)$ and $\bar{v}_2 = (0,0,-3,1)$.

In each of Problems 15-26 we give the reduced echelon form E of the coefficient matrix A, the parametric expressions for the variables, and the resulting basis vectors for the solution space.

15. $E = \begin{bmatrix} 1 & 0 & -11 \\ 0 & 1 & -7 \end{bmatrix}$ $\quad x_3 = t$, $x_2 = 7t$, $x_1 = 11t$

Basis vector $\bar{v}_1 = (11,7,1)$

16. $E = \begin{bmatrix} 1 & 0 & -11 \\ 0 & 1 & 5 \end{bmatrix}$ $x_3 = t$, $x_2 = -5t$, $x_1 = 11t$

Basis vector $\bar{v}_1 = (11,-5,1)$

17. $E = \begin{bmatrix} 1 & 0 & 11 & 11 \\ 0 & 1 & 3 & 5 \end{bmatrix}$ $x_3 = s$, $x_4 = t$
$x_2 = -3s - 5t$, $x_1 = -11s - 11t$

Basis vectors $\bar{v}_1 = (-11,-3,1,0)$, $\bar{v}_2 = (-11,-5,0,1)$

18. $E = \begin{bmatrix} 1 & 3 & 0 & 25 \\ 0 & 0 & 1 & -5 \end{bmatrix}$ $x_2 = s$, $x_4 = t$
$x_3 = 5t$, $x_1 = -3s - 25t$

Basis vectors $\bar{v}_1 = (-3,1,0,0)$, $\bar{v}_2 = (-25,0,5,0)$

19. $E = \begin{bmatrix} 1 & 0 & -3 & 4 \\ 0 & 1 & 2 & 3 \\ 0 & 0 & 0 & 0 \end{bmatrix}$ $x_3 = s$, $x_4 = t$
$x_2 = -2s - 3t$, $x_1 = 3s - 4t$

Basis vectors $\bar{v}_1 =(3,-2,1,0)$, $\bar{v}_2 = (-4,-3,0,1)$

20. $E = \begin{bmatrix} 1 & 0 & -1 & 2 \\ 0 & 1 & 3 & -1 \\ 0 & 0 & 0 & 0 \end{bmatrix}$ $x_3 = s$, $x_4 = t$
$x_2 = -3s + t$, $x_1 = s - 2t$

Basis vectors $\bar{v}_1 = (1,-3,1,0)$, $\bar{v}_2 = (-2,1,0,1)$

21. $E = \begin{bmatrix} 1 & 0 & 1 & 5 \\ 0 & 1 & 1 & 3 \\ 0 & 0 & 0 & 0 \end{bmatrix}$ $x_3 = s$, $x_4 = t$
$x_2 = -s - 3t$, $x_1 = -s - 5t$

Basis vectors $\bar{v}_1 = (-1,-1,1,0)$, $\bar{v}_2 = (-5,-3,0,1)$

22. $E = \begin{bmatrix} 1 & -2 & 0 & 5 \\ 0 & 0 & 1 & 7 \\ 0 & 0 & 0 & 0 \end{bmatrix}$ $x_2 = s$, $x_4 = t$
$x_3 = -7t$, $x_1 = 2s - 5t$

Basis vectors $\bar{v}_1 = (2,1,0,0)$, $\bar{v}_2 = (-5,0,-7,1)$

23. $E = \begin{bmatrix} 1 & 0 & -2 & 0 \\ 0 & 1 & 3 & 0 \\ 0 & 0 & 0 & 1 \end{bmatrix}$ $x_3 = t,\ x_4 = 0$
$x_2 = -3t,\ x_1 = 2t$

Basis vector $\bar{v}_1 = (2,-3,1,0)$

24. $E = \begin{bmatrix} 1 & 0 & 2 & 1 & 3 \\ 0 & 1 & -2 & -3 & 1 \\ 0 & 0 & 0 & 0 & 0 \end{bmatrix}$ $x_3 = r,\ x_4 = s,\ x_5 = t$
$x_2 = 2r + 3s - t$
$x_1 = -2r - s - 3t$

Basis vectors $\bar{v}_1 = (-2,2,1,0,0)$, $\bar{v}_2 = (-1,3,0,1,0)$
$\bar{v}_3 = (-3,-1,0,0,1)$

25. $E = \begin{bmatrix} 1 & 2 & 0 & -2 & 3 \\ 0 & 0 & 1 & -1 & 4 \\ 0 & 0 & 0 & 0 & 0 \end{bmatrix}$ $x_2 = r,\ x_4 = s,\ x_5 = t$
$x_3 = s - 4t$
$x_1 = -2r + 2s - 3t$

Basis vectors $\bar{v}_1 = (-2,1,0,0,0)$, $\bar{v}_2 = (2,0,1,1,0)$
$\bar{v}_3 = (-3,0,-4,0,1)$

26. $E = \begin{bmatrix} 1 & 0 & 0 & 2 & -3 \\ 0 & 1 & 0 & -1 & 4 \\ 0 & 0 & 1 & -2 & -5 \end{bmatrix}$ $x_4 = s,\ x_5 = t$
$x_3 = 2s + 5t,\ x_2 = s - 4t$
$x_1 = -2s + 3t$

Basis vectors $\bar{v}_1 = (-2,1,2,1,0)$, $\bar{v}_2 = (3,-4,5,0,1)$

27. If the vectors $\bar{v}_1, \bar{v}_2, \cdots, \bar{v}_n$ are linearly independent, and $\bar{w}$ is another vector in V, then the vectors $\bar{w}, \bar{v}_1, \bar{v}_2, \cdots, \bar{v}_n$ are linearly dependent. Hence there exist scalars $c, c_1, c_2, \cdots, c_n$ not all zero such that

$$c\bar{w} + c_1\bar{v}_1 + c_2\bar{v}_2 + \cdots + c_n\bar{v}_n = \bar{0}.$$

If $c = 0$ then the coefficients $c_1, c_2, \cdots, c_n$ would not all be zero, and so this equation would say that $\bar{v}_1, \bar{v}_2, \cdots, \bar{v}_n$ are linearly dependent. Therefore $c \neq 0$, so we can solve for $\bar{w}$ as a linear combination of $\bar{v}_1, \bar{v}_2, \cdots, \bar{v}_n$. Thus the linearly independent vectors $\bar{v}_1, \bar{v}_2, \cdots, \bar{v}_n$ span V, and hence form a basis for V.

28. If the n vectors in S were not linearly independent, then some one of them would be a linear combination of the others. These remaining n - 1 vectors would then span the n-dimensional vector space V, which is impossible. Thus the spanning set S is also linearly independent, and therefore is a basis for V.

29. Suppose that

$$c\bar{v} + c_1\bar{v}_1 + c_2\bar{v}_2 + \cdots + c_k\bar{v}_k = \bar{0}.$$

Then $c = 0$ because $\bar{v}$ is not in W. But then $c_1 = c_2 = \cdots = c_k = 0$ also because the vectors $\bar{v}_1, \bar{v}_2, \cdots, \bar{v}_k$ are linearly independent. Hence we have shown that our k+1 vectors are linearly independent.

30. Let $S = \{\bar{v}_1, \bar{v}_2, \cdots, \bar{v}_k\}$ be a linearly independent set of $k < n$ vectors in V. If $\bar{v}_{k+1}$ is not in W = span(S), then Problem 29 shows that the k+1 vectors $\bar{v}_1, \bar{v}_2, \cdots, \bar{v}_k, \bar{v}_{k+1}$ are linearly independent. Continuing in this fashion, we add one vector at a time until we have n linearly independent vectors in V, which then form a basis for B that contains S.

31. Obviously every linear combination of $\bar{v}_1, \bar{v}_2, \cdots, \bar{v}_k, \bar{v}_{k+1}$ is also a linear combination of $\bar{v}_1, \bar{v}_2, \cdots, \bar{v}_k$. But the former set of k+1 vectors spans V, so the latter set of k vectors also spans V.

32. If the spanning set S for V is not linearly independent, then some vector in S is a linear combination of the others. But Problem 31 says that when we remove this dependent vector from S, the resulting set of one fewer vectors still spans V. Continuing in this fashion, we remove one vector at a time from S until we get a spanning set that is also linearly independent, and is therefore a basis that is contained by the original spanning set S.

33. If S is a maximal linearly independent set in V, then we see immediately that every other vector in V is a linear combination of the vectors in S. Thus S also spans V, and is therefore a basis.

34. If the minimal spanning set S were not linearly independent, then some vector in S would be a linear combination of the others. Then the set obtained from S by deleting this dependent vector would be a smaller spanning set. Hence the spanning set S also is linearly independent, and therefore is a basis for V.

35. Let $S = \{\bar{v}_1, \bar{v}_2, \cdots, \bar{v}_n\}$ be a uniquely spanning set for V. Then the fact, that

$$\bar{0} = 0\bar{v}_1 + 0\bar{v}_2 + \cdots + 0\bar{v}_n$$

is the unique expression of the zero vector as a linear combination of the vectors in S, means that S is linearly independent. Hence S is a basis for V.

36. If $c_1, c_2, \cdots, c_k$ are scalars, then the linear combination

$$c_1\bar{v}_1 + c_2\bar{v}_2 + \cdots + c_k\bar{v}_k$$

is a vector of the form $(*, \cdots, *, c_1, c_2, \cdots, c_k)$. Hence this linear combination can equal the zero vector only if $c_1 = c_2 = \cdots = c_k = 0$. Thus the vectors $\bar{v}_1, \bar{v}_2, \cdots, \bar{v}_k$ are linearly independent.

SECTION 4.4

ROW AND COLUMN SPACES

Conventional wisdom (at a certain level) has it that a homogeneous linear system $A\bar{x} = \bar{0}$ of m equations in $n > m$ unknowns ought to have $n - m$ independent solutions. In Section 4.4 we use row and column spaces to show that this "conventional wisdom" is valid under the condition that the m equations are irredundant, meaning that the rank of the coefficient matrix A is m (so its row vectors are linearly independent).

In each of Problems 1-12 we give the reduced echelon form E of the matrix A, a basis for the row space of A, and a basis for the column space of A.

1. $E = \begin{bmatrix} 1 & 0 & 11 \\ 0 & 1 & -4 \\ 0 & 0 & 0 \end{bmatrix}$

Row basis: first and second row vectors of E
Column basis: first and second column vectors of A

2. $E = \begin{bmatrix} 1 & 0 & 2 \\ 0 & 1 & -3 \\ 0 & 0 & 0 \end{bmatrix}$

Row basis: first and second row vectors of E
Column basis: first and second column vectors of A

3. $E = \begin{bmatrix} 1 & 0 & 1 & 5 \\ 0 & 1 & 1 & 3 \\ 0 & 0 & 0 & 1 \end{bmatrix}$

Row basis: the first two row vectors of E
Column basis: the first two column vectors of A

4. $E = \begin{bmatrix} 1 & 0 & 0 & 4 \\ 0 & 1 & 0 & 3 \\ 0 & 0 & 1 & 0 \end{bmatrix}$

Row basis: the three row vectors of E
Column basis: the first three column vectors of A

5. $E = \begin{bmatrix} 1 & 0 & -2 & 0 \\ 0 & 1 & 3 & 0 \\ 0 & 0 & 0 & 1 \end{bmatrix}$

Row basis: the three row vectors of E

Column basis: the first, second, and fourth column vectors of A

6. $E = \begin{bmatrix} 1 & 0 & 1 & 0 \\ 0 & 1 & 2 & 0 \\ 0 & 0 & 0 & 1 \end{bmatrix}$

Row basis: the three row vectors of E

Column basis: the first, second, and fourth column vectors of A

7. $E = \begin{bmatrix} 1 & 0 & -3 & 4 \\ 0 & 1 & 2 & 3 \\ 0 & 0 & 0 & 0 \\ 0 & 0 & 0 & 0 \end{bmatrix}$

Row basis: the first two row vectors of E

Column basis: the first two column vectors of A

8. $E = \begin{bmatrix} 1 & 0 & 1 & 0 \\ 0 & 1 & 2 & 0 \\ 0 & 0 & 0 & 1 \\ 0 & 0 & 0 & 0 \end{bmatrix}$

Row basis: the first three row vectors of E

Column basis: the first, second, and fourth column vectors of A

9. $E = \begin{bmatrix} 1 & 0 & 0 & 3 \\ 0 & 1 & 0 & -2 \\ 0 & 0 & 1 & 4 \\ 0 & 0 & 0 & 0 \end{bmatrix}$

Row basis: the first three row vectors of E

Column basis: the first three column vectors of A

10. $E = \begin{bmatrix} 1 & 0 & 1 & 0 & 0 \\ 0 & 1 & 1 & 0 & 1 \\ 0 & 0 & 0 & 1 & 1 \\ 0 & 0 & 0 & 0 & 0 \end{bmatrix}$

Row basis: the first three row vectors of E

Column basis: the first, second, and fourth column vectors of A

11. $E = \begin{bmatrix} 1 & 0 & 2 & 1 & 0 \\ 0 & 1 & 1 & 2 & 0 \\ 0 & 0 & 0 & 0 & 1 \\ 0 & 0 & 0 & 0 & 0 \end{bmatrix}$

Row basis: the first three row vectors of E

Column basis: the first, second, and fifth column vectors of A

12. E = same echelon matrix as in Problem 11

Row basis: the first three row vectors of E

Column basis: the first, second, and fifth column vectors of A

In each of Problems 13-16 we give the reduced echelon form E of the matrix having the given vectors as its column vectors.

13. $E = \begin{bmatrix} 1 & 0 & 1 \\ 0 & 1 & 2 \\ 0 & 0 & 0 \\ 0 & 0 & 0 \end{bmatrix}$

Linearly independent: $\bar{v}_1$ and $\bar{v}_2$

14. $E = \begin{bmatrix} 1 & 0 & 1/5 & 2 \\ 0 & 1 & 2/5 & 1 \\ 0 & 0 & 0 & 0 \\ 0 & 0 & 0 & 0 \end{bmatrix}$

Linearly independent: $\bar{v}_1$ and $\bar{v}_2$

15. $E = \begin{bmatrix} 1 & 0 & 2 & 0 \\ 0 & 1 & -1 & 0 \\ 0 & 0 & 0 & 1 \\ 0 & 0 & 0 & 0 \end{bmatrix}$

Linearly independent: $\bar{v}_1$, $\bar{v}_2$, and $\bar{v}_4$

16. $E = \begin{bmatrix} 1 & 0 & 2 & 0 & 0 \\ 0 & 1 & -1 & 0 & 0 \\ 0 & 0 & 0 & 1 & 0 \\ 0 & 0 & 0 & 0 & 1 \end{bmatrix}$

Linearly independent: $\bar{v}_1, \bar{v}_2, \bar{v}_4$, and $\bar{v}_5$

In each of Problems 17-20 the matrix E is the reduced echelon form of the matrix

$$A = [\bar{v}_1 \cdots \bar{v}_k \ \bar{e}_1 \cdots \bar{e}_n].$$

17. $E = \begin{bmatrix} 1 & 0 & -3 & 0 & 2 \\ 0 & 1 & 2 & 0 & -1 \\ 0 & 0 & 0 & 1 & -1 \end{bmatrix}$

Basis: $\bar{v}_1, \bar{v}_2, \bar{e}_2$

18. $E = \begin{bmatrix} 1 & 0 & 1/5 & 0 & -2/5 \\ 0 & 1 & 1/5 & 0 & 3/5 \\ 0 & 0 & 0 & 1 & 2 \end{bmatrix}$

Basis: $\bar{v}_1, \bar{v}_2, \bar{e}_2$

19. $E = \begin{bmatrix} 1 & 0 & 3 & 0 & 0 & -2 \\ 0 & 1 & -1 & 0 & 0 & 1 \\ 0 & 0 & 0 & 1 & 0 & -1 \\ 0 & 0 & 0 & 0 & 1 & -1 \end{bmatrix}$

Basis: $\bar{v}_1, \bar{v}_2, \bar{e}_2, \bar{e}_3$

20. $E = \begin{bmatrix} 1 & 0 & 0 & -5/2 & 0 & 2 \\ 0 & 1 & 0 & 3/2 & 0 & -1 \\ 0 & 0 & 1 & 0 & 0 & -1 \\ 0 & 0 & 0 & 0 & 1 & -1 \end{bmatrix}$

Basis: $\bar{v}_1, \bar{v}_2, \bar{e}_1, \bar{e}_3$

In each of Problems 21-24 the matrix E is the <u>reduced</u> echelon form of the transpose A^T of the coefficient matrix A.

21. $E = \begin{bmatrix} 1 & 0 & 2 \\ 0 & 1 & 1 \\ 0 & 0 & 0 \end{bmatrix}$

The first and second equations are irredundant.

22. $E = \begin{bmatrix} 1 & 0 & 2 \\ 0 & 1 & 1 \\ 0 & 0 & 0 \\ 0 & 0 & 0 \end{bmatrix}$

The first and second equations are irredundant.

23. $E = \begin{bmatrix} 1 & 0 & 2 & 0 \\ 0 & 1 & 1 & 0 \\ 0 & 0 & 0 & 1 \\ 0 & 0 & 0 & 0 \end{bmatrix}$

The first, second, and fourth equations are irredundant.

24. $E = \begin{bmatrix} 1 & 0 & 1 & 2 & 0 \\ 0 & 1 & 2 & 1 & 0 \\ 0 & 0 & 0 & 0 & 1 \end{bmatrix}$

The first, second, and fifth equations are irredundant.

25. The row vectors of A are the column vectors of A^T, so

$$\begin{aligned} \text{rank}(A) &= \text{row rank of } A \\ &= \text{column rank of } A^T = \text{rank}(A^T). \end{aligned}$$

26. The rank of the $n \times n$ matrix A is n if and only if its column vectors are linearly independent, in which case $\det A \neq 0$ by Theorem 2 in Section 4.2, so A is invertible by Theorem 2 in Section 2.3.

27. The rank of the 3×5 matrix A is 3, so its column vectors span R^3. Therefore any vector $\bar{b}$ in R^3 can be expressed as a linear combination of the column vectors of A. This gives a solution of $A\bar{x} = \bar{b}$.

28. The rank of the 5×3 matrix A is 3, so its three column vectors are linearly independent. Therefore a vector $\bar{b}$ in R^5 can be expressed in at most one way as a linear combination of the column vectors of A. This means that $A\bar{x} = \bar{b}$ has at most one solution.

29. The rank of A is at most m, and therefore is less than n. Hence the column vectors of A are linearly dependent. It follows that, if $\bar{b}$ can be expressed as a linear combination of the column vectors of A, then this can be done in many different ways.

30. The rank of A is at most n, which is less than m. Hence the column space of A is a proper subspace of R^m, so there exists a vector $\bar{b}$ in R^m that is not a linear combination of the column vectors of A. This means that $A\bar{x} = \bar{b}$ has no solution.

31. The rank of the $m \times n$ matrix A is m if and only if A has m linearly independent column vectors that span R^m, so

every vector $\bar{b}$ in R^m can be expressed as a linear combination of the column vectors of A. This means that $A\bar{x} = \bar{b}$ has a solution.

32. The rank of the $m \times n$ matrix A is n if and only if the n column vectors of A are linearly independent, in which case a vector $\bar{b}$ in R^m can be expressed in at most one way as a linear combination of the column vectors of A. This means that $A\bar{x} = \bar{b}$ has at most one solution.

33. Suppose that some linear combination of the k pivot column vectors in (8) equals the zero vector. If the coefficients in this linear combination are $c_1, c_2, \cdots, c_k$, then by examining components we find that

$$\begin{aligned} c_1d_1 + \cdots\cdots\cdots\cdots\cdots\cdots &= 0 \\ c_2d_2 + \cdots\cdots\cdots\cdots\cdots &= 0 \\ c_3d_3 + \cdots\cdots\cdots\cdots &= 0 \\ &\vdots \\ c_kd_k &= 0. \end{aligned}$$

Because the leading entries $d_1, d_2, \cdots, d_k$ are all nonzero, it follows that $c_1 = c_2 = \cdots = c_k = 0$. Thus the k pivot column vectors are linearly independent.

34. If no row interchanges are involved, then (for any k) the space spanned by the first k row vectors of A is never changed in the process of reducing A to the echelon matrix E; this follows immediately from the proof of Theorem 2. Hence the first r row vectors of A span the r-dimensional space Row(A), and therefore are linearly independent.

36. Look at the r row vectors of A determined by the largest nonsingular $r \times r$ submatrix. Then Theorem 3 in Section 4.2 says that these r row vectors are linearly independent, whereas any r+1 row vectors of A are linearly dependent.

SECTION 4.5

GENERAL VECTOR SPACES

This optional section consists largely of examples of vector spaces of matrices and functions. Students who have not studied calculus can read all of the section through Example 6. Such students should omit Problems 23-30.

1. Yes; any linear combination of diagonal matrices obviously is a diagonal matrix.

2. Yes; any linear combination of symmetric matrices obviously is a symmetric matrix.

3. No; it does not contain the zero matrix.

4. No; the sum

$$\begin{bmatrix} 1 & 0 & 0 \\ 0 & 0 & 0 \\ 0 & 0 & 0 \end{bmatrix} + \begin{bmatrix} 0 & 0 & 0 \\ 0 & 1 & 0 \\ 0 & 0 & 1 \end{bmatrix} = \begin{bmatrix} 1 & 0 & 0 \\ 0 & 1 & 0 \\ 0 & 0 & 1 \end{bmatrix}$$

of singular matrices is nonsingular.

5. Yes; if $h = af + bg$ and $f(0) = g(0) = 0$, then $h(0) = af(0) + bg(0) = 0$.

6. No; it does not contain the zero function.

7. No; for the same reason as in Problem 6.

8. Yes, because if $f(-x) = -f(x)$ and $g(-x) = -g(x)$ then

$$\begin{aligned}(af + bg)(-x) &= af(-x) + bg(-x) \\ &= -af(x) - bg(x) \\ &= -(af + bg)(x).\end{aligned}$$

9. No; it does not contain the zero polynomial.

10. Yes; any linear combination of polynomials of the form

$a_2x^2 + a_3x^3$ is a polynomial of this form.

11. Yes, because $c \cdot 0 + d \cdot 0 = 0$.

12. No, because a real multiple of an integer is not always an integer.

13. Yes; obviously neither is a scalar multiple of the other.

14. Yes; obviously neither is a scalar multiple of the other.

15. If

$$c_1(1 + x) + c_2(1 - x) + c_3(1 - x^2) = 0$$

then

$$c_1 + c_2 + c_3 = 0, \quad c_1 - c_2 = 0, \text{ and } \quad -c_3 = 0.$$

It follows easily that $c_1 = c_2 = c_3 = 0$, so the three polynomials are linearly independent.

16. $(-1)(1 + x) + (1)(x + x^2) + (1)(1 - x^2) = 0$, so the three polynomials are linearly dependent.

17. $\cos 2x = \cos^2 x - \sin^2 x$, so the three functions are linearly dependent.

18. If

$$c_1(2 \cos x + 3 \sin x) + c_2(4 \cos x + 5 \sin x) = 0$$

then the fact that $\cos x$ and $\sin x$ are linearly independent implies that

$$2c_1 + 4c_2 = 3c_1 + 5c_2 = 0.$$

It follows easily that $c_1 = c_2 = 0$, so the original two functions are linearly independent.

19. Multiplication by $(x - 2)(x - 3)$ yields

$$x - 5 = A(x - 3) + B(x - 2) = (A + B)x - (3A + 2B).$$

Hence $A + B = 1$ and $3A + 2B = 5$, and it follows that $A = 3$ and $B = -2$.

20. Multiplication by $x(x^2 - 1)$ yields

$$2 = A(x - 1)(x + 1) + Bx(x + 1) + Cx(x - 1)$$

$$= (A + B + C)x^2 + (B - C)x + (-A)$$

Hence $A + B + C = 0$, $B - C = 0$, and $-A = 2$. It follows readily that $A = -2$ and $B = C = 1$.

21. Multiplication by $x(x^2 + 4)$ yields

$$8 = A(x^2 + 4) + Bx^2 + Cx.$$

Hence $A + B = 0$, $C = 0$, and $4A = 8$, so it follows that $A = 2$ and $B = -2$.

22. Multiplication by $(x + 1)(x + 2)(x + 3)$ yields

$$2x = A(x + 2)(x + 3) + B(x + 1)(x + 3) + C(x + 1)(x + 2)$$

$$= (A + B + C)x^2 + (5A + 4B + 3C)x + (6A + 3B + 2C).$$

Hence

$$\begin{aligned} A + B + C &= 0 \\ 5A + 4B + 3C &= 2 \\ 6A + 3B + 2C &= 0, \end{aligned}$$

and it follows that $A = -1$, $B = 4$, $C = -3$.

23. If $y(x) = Ae^x + Be^{2x}$ then the equations

$$\begin{aligned} y(0) &= A + B = 1 \\ y'(0) &= A + 2B = -1 \end{aligned}$$

yield $A = 3$ and $B = -2$. Hence $y(x) = 3e^x - 2e^{2x}$.

24. If $y(x) = Ae^{3x} + Be^{-2x}$ then the equations

$$y(0) = A + B = 7$$
$$y'(0) = 3A - 2B = 11$$

yield $A = 5$ and $B = 2$. Hence $y(x) = 5e^{3x} + 2e^{-2x}$.

25. If $y(x) = A \cos 2x + B \sin 2x$ then the equations

$$y(0) = A = 4, \quad y'(0) = 2B = 6$$

yield $y(x) = 4 \cos 2x + 3 \sin 2x$.

26. If $y(x) = Ae^x + Bxe^x$ then

$$y'(x) = (A + B)e^x + Bxe^x.$$

Hence $y(0) = A = 5$ and $y'(0) = A + B = 2$, so $B = -3$. Therefore $y(x) = 5e^x - 3xe^x$.

27. (a) The functions e^x and e^{-x} obviously are linearly independent, so they form a basis for the 2-dimensional solution space.

(b) If

$$0 = A \cosh x + B \sinh x$$
$$= (1/2)(A + B)e^x + (1/2)(A - B)e^{-x}$$

then $A + B = A - B = 0$. It follows that $A = B = 0$, so the functions $\cosh x$ and $\sinh x$ are linearly independent.

28. If $W(x) \neq 0$ for some x, then for that x the homogeneous linear system

$$c_1f(x) + c_2g(x) + c_3h(x) = 0$$
$$c_1f'(x) + c_2g'(x) + c_3h'(x) = 0$$
$$c_1f''(x) + c_2g''(x) + c_3h''(x) = 0$$

with nonsingular coefficient matrix has only the trivial solution $c_1 = c_2 = c_3 = 0$.

29. (a)

$$W(x) = \begin{vmatrix} e^x & e^{-x} & e^{2x} \\ e^x & -e^{-x} & 2e^{2x} \\ e^x & e^{-x} & 4e^{2x} \end{vmatrix} = e^{2x} \begin{vmatrix} 1 & 1 & 1 \\ 1 & -1 & 2 \\ 1 & 1 & 4 \end{vmatrix} = -6e^{2x} \neq 0$$

(b) If $y(x) = Ae^x + Be^{-x} + Ce^{2x}$ then the equations

$$\begin{aligned} y(0) &= A + B + C = 0 \\ y'(0) &= A - B + 2C = 7 \\ y''(0) &= A + B + 4C = 3 \end{aligned}$$

yield $A = 2$, $B = -3$, and $C = 1$. Therefore

$$y(x) = 2e^x - 3e^{-x} + e^{2x}.$$

30. (a)

$$W(x) = \begin{vmatrix} e^x & \cos 2x & \sin 2x \\ e^x & -2\sin 2x & 2\cos 2x \\ e^x & -4\cos 2x & -4\sin 2x \end{vmatrix} = 10e^x \neq 0$$

(b) If $y(x) = Ae^x + B \cos 2x + C \sin 2x$ then the equations

$$\begin{aligned} A + B \quad\quad &= 1 \\ A \quad\quad + 2C &= 5 \\ A - 4B \quad\quad &= -5 \end{aligned}$$

yield $A = 3$, $B = -2$, and $C = 1$. Therefore

$$y(x) = 3e^x - 2 \cos 2x + \sin 2x.$$

31. (a) The verification in a componentwise manner that V is a vector space is the same as the verification that R^n is a vector space, except with vectors having infinitely many rather than finitely many components.

(b) If $e_n = \{0, \cdots, 0, 1, 0, \cdots\}$ is the sequence with 1 in the nth position, then any finite set of the vectors $e_1, e_2, e_3, \cdots$ is linearly independent.

32. (a) Obviously any linear combination of two such sequences is such a sequence.

(b) Let $v_1 = \{1,0,1,1,2,3,5,\cdots\}$ be the element with $x_1 = 1$ and $x_2 = 0$, and let $v_2 = \{0,1,1,2,3,5,\cdots\}$ be the element with $x_1 = 0$ and $x_2 = 1$. Then v_1 and v_2 form a basis for W.

33. (a) If $z_1 = a_1 + b_1 i$ and $z_2 = a_2 + b_2 i$, then it is obvious that

$$T(c_1 z_1 + c_2 z_2) = c_1 T(z_1) + c_2 T(z_2)$$

when each side is expressed in terms of $a_1, a_2, b_1, b_2, c_1, c_2$.

(b) This follows upon comparison of Equation (5) and the preceding formula in the text.

(c) If $z = a + bi$ then

$$\frac{1}{z} = \frac{1}{a + bi} \cdot \frac{a - bi}{a - bi} = \frac{a - bi}{a^2 + b^2} .$$

Therefore

$$T(z^{-1}) = \frac{1}{a^2 + b^2} \begin{bmatrix} a & -b \\ b & a \end{bmatrix}$$

$$= \begin{bmatrix} a & b \\ -b & a \end{bmatrix}^{-1} = T(z)^{-1}.$$

CHAPTER 5

ORTHOGONALITY AND LEAST SQUARES

At the beginning of this chapter it may be a good idea to let students know that the theory of orthogonality and orthogonal projections in Sections 5.1 and 5.2 is not developed "just for its own sake". To the contrary, orthogonality underlies the least squares solutions of Section 5.2 and the curve fitting methods of Section 5.3 that are among the most practical and important of all the applications of linear algebra.

SECTION 5.1

ORTHOGONAL VECTORS IN R^n

In this section the dot product is generalized to vectors in R^n; more general inner products are deferred to Section 5.5. The dot product enables us to flesh out the algebra of vectors in R^n with the Euclidean geometry of angles and distance; we can now refer to the vector space R^n (provided with the dot product) as n-dimensional Euclidean space. It should be emphasized that the material on orthogonal complements is not merely an addendum to the discussion of orthogonality; it is an essential prerequisite to the subsequent sections of Chapter 5.

In each of Problems 1 through 4 we calculate the dot product of each two of the three given vectors and find that

$$\bar{v}_1\cdot\bar{v}_2 = \bar{v}_1\cdot\bar{v}_3 = \bar{v}_2\cdot\bar{v}_3 = 0.$$

Hence the three vectors are mutually orthogonal.

In each of Problems 5-8 we write $\bar{u} = \overrightarrow{CB}$, $\bar{v} = \overrightarrow{CA}$, and $\bar{w} = \overrightarrow{AB}$. Then we calculate $a = |\bar{u}|$, $b = |\bar{v}|$, and $c = |\bar{w}|$ so as to verify that $a^2 + b^2 = c^2$.

5. $\bar{u} = (1,1,2,-1)$, $\bar{v} = (1,-1,1,2)$, $\bar{w} = (0,2,1,-3)$;

$a^2 = 7$, $b^2 = 7$, $c^2 = 14$

6. $\bar{u} = (3,-1,2,2)$, $\bar{v} = (2,2,-3,1)$, $\bar{w} = (1,-3,5,1)$;

$a^2 = 18$, $b^2 = 18$, $c^2 = 36$

7. $\bar{u} = (2,1,-2,1,3)$, $\bar{v} = (3,2,2,2,-2)$, $\bar{w} = (-1,-1,-4,-1,5)$;

$a^2 = 19$, $b^2 = 25$, $c^2 = 44$

8. $\bar{u} = (3,2,4,5,7)$, $\bar{v} = (7,5,-5,2,-3)$, $\bar{w} = (-4,-3,9,3,10)$;

$a^2 = 103$, $b^2 = 112$, $c^2 = 215$

The computations in Problems 5-8 show that in each triangle the angle at C is a right angle. The angles at the vertices A and B are determined by

$$\cos A = \frac{\bar{u}\cdot\bar{w}}{|u|\,|w|}, \qquad \cos B = \frac{\bar{v}\cdot\bar{w}}{|v|\,|w|} .$$

9. $A = B = 45^\circ$

10. $A = B = 45^\circ$

11. $A \approx 48.92^\circ$, $B \approx 41.08^\circ$

12. $A \approx 46.20^\circ$, $B \approx 43.80^\circ$

In each of Problems 13-22, we denote by A the matrix having the given vectors as its row vectors, and by E the reduced echelon form of A. From E we find the general solution of the system $A\bar{x} = \bar{0}$ in terms of parameters $s,t,\cdots$. We then get basis vectors $\bar{u}_1,\bar{u}_2,\cdots$ for the orthogonal complement of V by setting each parameter in turn equal to 1 (and the others equal to 0).

13. $A = E = [1 \quad -2 \quad 3]$

$x_2 = s$, $x_3 = t$, $x_1 = 2s - 3t$

$\bar{u}_1 = (2,1,0)$, $\bar{u}_2 = (-3,0,1)$

14. $A = E = [1 \quad 5 \quad -3]$

$x_2 = s$, $x_3 = t$, $x_1 = -5s + 3t$

$\bar{u}_1 = (-5,1,0)$, $\bar{u}_2 = (3,0,1)$

15. $A = E = [1 \quad -2 \quad -3 \quad 5]$

$x_2 = r$, $x_3 = s$, $x_4 = t$, $x_1 = 2r + 3s - 5t$

$\bar{u}_1 = (2,1,0,0)$, $\bar{u}_2 = (3,0,1,0)$, $\bar{u}_3 = (-5,0,0,1)$

16. $A = E = [1 \quad 7 \quad -6 \quad -9]$

$x_2 = r$, $x_3 = s$, $x_4 = t$, $x_1 = -7r + 6s + 9t$

$\bar{u}_1 = (-7,1,0,0)$, $\bar{u}_2 = (6,0,1,0)$, $\bar{u}_3 = (9,0,0,1)$

17. $E = \begin{bmatrix} 1 & 0 & -7 & 19 \\ 0 & 1 & 3 & -5 \end{bmatrix}$

$x_3 = s$, $x_4 = t$, $x_2 = -3s + 5t$, $x_1 = 7s - 19t$

$\bar{u}_1 = (7,-3,1,0)$, $\bar{u}_2 = (-19,5,0,1)$

18. $E = \begin{bmatrix} 1 & 0 & 12 & -16 \\ 0 & 1 & 3 & -7 \end{bmatrix}$

$x_3 = s$, $x_4 = t$, $x_2 = -3s + 7t$, $x_1 = -12s + 16t$

$\bar{u}_1 = (-12,-3,1,0)$, $\bar{u}_2 = (16,7,0,1)$

19. $E = \begin{bmatrix} 1 & 0 & 13 & -4 & 11 \\ 0 & 1 & -4 & 3 & -4 \end{bmatrix}$

$x_3 = r, \quad x_4 = s, \quad x_5 = t,$

$x_2 = 4r - 3s + 4t, \quad x_1 = -13r + 4s - 11t$

$\bar{u}_1 = (-13,4,1,0,0), \quad \bar{u}_2 = (4,-3,0,1,0), \quad \bar{u}_3 = (-11,4,0,0,1)$

20. $E = \begin{bmatrix} 1 & 0 & 5 & 12 & 19 \\ 0 & 1 & -1 & -4 & -7 \end{bmatrix}$

$x_3 = r, \quad x_4 = s, \quad x_5 = t$

$x_2 = r + 4s + 7t, \quad x_1 = -5r - 12s - 19t$

$\bar{u}_1 = (-5,1,1,0,0), \quad \bar{u}_2 = (-12,4,0,1,0), \quad \bar{u}_3 = (-19,7,0,0,1)$

21. $E = \begin{bmatrix} 1 & 0 & 1 & 0 & 0 \\ 0 & 1 & 1 & 0 & 1 \\ 0 & 0 & 0 & 1 & 1 \end{bmatrix}$

$x_3 = s, \quad x_5 = t, \quad x_4 = -t, \quad x_2 = -s - t, \quad x_1 = -s$

$\bar{u}_1 = (-1,-1,1,0,0), \quad \bar{u}_2 = (0,-1,0,-1,1)$

22. $E = \begin{bmatrix} 1 & 0 & 2 & -1 & 0 \\ 0 & 1 & -1 & 2 & 0 \\ 0 & 0 & 0 & 0 & 1 \end{bmatrix}$

$x_3 = s, \quad x_4 = t, \quad x_5 = 0, \quad x_2 = s - 2t, \quad x_1 = -2s + t$

$\bar{u}_1 = (-2,1,1,0,0), \quad \bar{u}_2 = (1,-2,0,1,0)$

23. (a) $|\bar{u} + \bar{v}|^2 + |\bar{u} - \bar{v}|^2$

$$= (\bar{u}\cdot\bar{u} + 2\bar{u}\cdot\bar{v} + \bar{v}\cdot\bar{v}) + (\bar{u}\cdot\bar{u} - 2\bar{u}\cdot\bar{v} + \bar{v}\cdot\bar{v})$$

$$= 2\bar{u}\cdot\bar{u} + 2\bar{v}\cdot\bar{v}$$

$$= 2|\bar{u}|^2 + 2|\bar{v}|^2$$

(b) Similar to (a).

24. Equation (15) in the text says that the given formula holds for $k = 2$ vectors. Assume inductively that it holds for

$k = n - 1$ vectors. Then

$$|\bar{v}_1 + \bar{v}_2 + \cdots + \bar{v}_{n-1} + \bar{v}_n|^2$$

$$= |\bar{v}_1 + \bar{v}_2 + \cdots + \bar{v}_{n-1}|^2 + |\bar{v}_n|^2$$

$$= |\bar{v}_1|^2 + |\bar{v}_2|^2 + \cdots + |\bar{v}_n|^2$$

as desired.

25. Suppose, for instance, that $A = \bar{e}_1 = (1,0,0,0,0)$, $B = \bar{e}_3 = (0,0,1,0,0)$, and $C = \bar{e}_5 = (0,0,0,0,1)$ in R^5. Then $\bar{e}_1 - \bar{e}_3 = (1,0,-1,0,0)$ and $\bar{e}_1 - \bar{e}_5 = (1,0,0,0,-1)$. The dot product of the latter two vectors is 1, and the length of each is $\sqrt{2}$. Hence $\cos\theta = 1/2$, so $\theta = 60^\circ$.

26. Because $\bar{u}\cdot\bar{v} = |\bar{u}||\bar{v}|\cos\theta$, it follows that $|\bar{u}\cdot\bar{v}| = |\bar{u}||\bar{v}|$ if and only if $\cos\theta = \pm 1$, that is, if and only if either $\theta = 0$ or $\theta = \pi$. In either case $\bar{u}$ and $\bar{v}$ are collinear.

27. If $\bar{u}$ lies both in the subspace V and in its orthogonal complement, then the vector $\bar{u}$ is orthogonal to itself. Hence $\bar{u}\cdot\bar{u} = |\bar{u}|^2 = 0$, so $\bar{u} = \bar{0}$.

28. If W is the orthogonal complement of V, then every vector in V is orthogonal to every vector in W. Hence V is contained in the orthogonal complement of W. But it follows easily from Equation (18) that the latter two subspaces have the same dimension. Since one contains the other, they must actually be the same subspace of R^n.

29. If $\bar{u}$ is orthogonal to each vector in S, then it follows easily that $\bar{u}$ is orthogonal to any linear combination of vectors in S. Thus $\bar{u}$ is orthogonal to $V = \text{Span}(S)$.

30. If $\bar{u}\cdot\bar{v} = 0$ and $\bar{u} + \bar{v} = \bar{0}$ then

$$0 = \bar{u}\cdot(\bar{u} + \bar{v}) = \bar{u}\cdot\bar{u} + \bar{u}\cdot\bar{v} = \bar{u}\cdot\bar{u},$$

so $\bar{u} = \bar{0}$. Then $\bar{u} + \bar{v} = 0$ implies that $\bar{v} = \bar{0}$.

31. We want to show that any linear combination of vectors in S is orthogonal to every linear combination of vectors in T. But if each $\bar{u}_i$ is orthogonal to each $\bar{v}_j$, then it follows readily upon expanding that

$$(a_1\bar{u}_1 + \cdots + a_p\bar{u}_p)\cdot(b_1\bar{v}_1 + \cdots + b_q\bar{v}_q) = 0.$$

32. Suppose that

$$a_1\bar{u}_1 + a_2\bar{u}_2 + b_1\bar{v}_1 + b_2\bar{v}_2 = \bar{0},$$

and write

$$\bar{u} = a_1\bar{u}_1 + a_2\bar{u}_2, \quad \bar{v} = b_1\bar{v}_1 + b_2\bar{v}_2.$$

Then $\bar{u}$ and $\bar{v}$ are orthogonal by Problem 31, so by Problem 30 the fact that $\bar{u} + \bar{v} = \bar{0}$ implies that

$$\bar{u} = a_1\bar{u}_1 + a_2\bar{u}_2 = \bar{0}, \quad \bar{v} = b_1\bar{v}_1 + b_2\bar{v}_2 = \bar{0}.$$

Now the assumed linear independence of $\{\bar{u}_1, \bar{u}_2\}$ and of $\{\bar{v}_1, \bar{v}_2\}$ implies that $a_1 = a_2 = b_1 = b_2 = 0$, so the four vectors are linearly independent.

33. This is the same as Problem 32, except with

$$\bar{u} = a_1\bar{u}_1 + a_2\bar{u}_2 + \cdots + a_k\bar{u}_k$$

and

$$\bar{v} = b_1\bar{v}_1 + b_2\bar{v}_2 + \cdots + b_m\bar{v}_m.$$

34. It follows immediately from Problem 33 and from Equation (18) that the union of a basis for V and a basis for its orthogonal complement is a linearly independent set of n vectors, and is therefore a basis for R^n.

35. This is one of the important theorems of linear algebra. The nonhomogeneous system

$$A\bar{x} = \bar{b}$$

is consistent if and only if the vector $\bar{b}$ is in Col(A) = Row(A^T). But $\bar{b}$ is in Row(A^T) if and only if $\bar{b}$ is orthogonal to the orthogonal complement of Row(A^T), which is Null(A^T), the solution space of the homogeneous system

$$A^T\bar{y} = \bar{0}.$$

Thus we have proved that the nonhomogeneous system $A\bar{x} = \bar{b}$ has a solution if and only if $\bar{b}$ is orthogonal to every solution $\bar{y}$ of the homogeneous system $A^T\bar{y} = \bar{0}$.

SECTION 5.2

ORTHOGONAL PROJECTIONS AND LEAST SQUARES SOLUTIONS

In each of Problems 1-6 we apply the formula

$$\bar{p} = \frac{\bar{a}\cdot\bar{b}}{\bar{a}\cdot\bar{a}}\,\bar{a}$$

derived in Example 2.

1. $\bar{p} = 5\bar{a} = (5,5,5)$
2. $\bar{p} = 3\bar{a} = (2,4,6)$
3. $\bar{p} = 4\bar{a} = (4,0,4,4)$
4. $\bar{p} = 2\bar{a} = (4,2,0,-4)$
5. $\bar{p} = 2\bar{a} = (2,-2,2,4,-6)$
6. $\bar{p} = 3\bar{a} = (12,15,-15,9,3,6)$

In each of problems 7-16 we list the normal system

$$A^TA\bar{x} = A^T\bar{b},$$

the least squares solution $\bar{x}$, and the orthogonal projection $\bar{p} = A\bar{x}$.

7. $\begin{bmatrix} 6 & -2 \\ -2 & 3 \end{bmatrix} \bar{x} = \begin{bmatrix} 8 \\ 0 \end{bmatrix}$

   $\bar{x} = (12/7,8/7) \quad \bar{p} = (20/7,4/7,16/7)$

8. $\begin{bmatrix} 11 & 6 \\ 6 & 6 \end{bmatrix} \bar{x} = \begin{bmatrix} 13 \\ 9 \end{bmatrix}$

$\bar{x} = (4/5, 7/10)$ $\bar{p} = (11/5, 3/2, 31/10)$

9. $\begin{bmatrix} 11 & -3 \\ -3 & 3 \end{bmatrix} \bar{x} = \begin{bmatrix} 27 \\ -5 \end{bmatrix}$

$\bar{x} = (11/4, 13/12)$ $\bar{p} = (23/6, 5/3, 43/6)$

10. $\begin{bmatrix} 6 & -3 \\ -3 & 6 \end{bmatrix} \bar{x} = \begin{bmatrix} 10 \\ -2 \end{bmatrix}$

$\bar{x} = (2, 2/3)$ $\bar{p} = (8/3, 10/3, 2/3)$

11. $\begin{bmatrix} 7 & 4 \\ 4 & 7 \end{bmatrix} \bar{x} = \begin{bmatrix} 18 \\ 15 \end{bmatrix}$

$\bar{x} = (2, 1)$ $\bar{p} = (3, 1, 5, 4)$

12. $\begin{bmatrix} 12 & 8 \\ 8 & 10 \end{bmatrix} \bar{x} = \begin{bmatrix} 32 \\ 28 \end{bmatrix}$

$\bar{x} = (12/7, 10/7)$ $\bar{p} = (32/7, 2/7, 22/7, 8)$

13. $\begin{bmatrix} 6 & -3 & 0 \\ -3 & 6 & -5 \\ 0 & -5 & 6 \end{bmatrix} \bar{x} = \begin{bmatrix} 6 \\ 0 \\ -3 \end{bmatrix}$

$\bar{x} = (7/4, 3/2, 3/4)$ $\bar{p} = (1/4, 11/4, 1/4, 3/4)$

14. $\begin{bmatrix} 11 & -9 & -1 \\ -9 & 10 & 8 \\ -1 & 8 & 20 \end{bmatrix} \bar{x} = \begin{bmatrix} 4 \\ 4 \\ 20 \end{bmatrix}$

$\bar{x} = (-4/5, -8/5, 8/5)$ $\quad$ $\bar{p} = (-4/5, -12/5, 12/4, 16/5)$

15. $\begin{bmatrix} 3 & 1 & -3 \\ 1 & 6 & -4 \\ -3 & -4 & 9 \end{bmatrix} \bar{x} = \begin{bmatrix} 5 \\ -10 \\ 10 \end{bmatrix}$

$\bar{x} = (4, -1, 2)$ $\quad$ $\bar{p} = (5, -5, 0, 0)$

Thus $\bar{p} = \bar{b}$, so $\bar{b}$ is in the column space of A.

16. $\begin{bmatrix} 19 & -1 & -2 \\ -1 & 3 & -4 \\ -2 & -4 & 7 \end{bmatrix} \bar{x} = \begin{bmatrix} 80 \\ 0 \\ -15 \end{bmatrix}$

$\bar{x} = (25/6, 1/2, -2/3)$ $\quad$ $\bar{p} = (3, -2/3, 43/3, 34/3)$

In each of Problems 17-22 we list the column vectors $\bar{a}_1, \bar{a}_2, \cdots$ of A that form a basis for V, the normal system $A^T A\bar{x} = A^T\bar{b}$, the least squares solution $\bar{x}$, and the orthogonal projection $\bar{p} = A\bar{x}$.

17. $\bar{a}_1 = (3, 1, 0)$, $\bar{a}_2 = (-2, 0, 1)$

$\begin{bmatrix} 10 & -6 \\ -6 & 5 \end{bmatrix} \bar{x} = \begin{bmatrix} 56 \\ 0 \end{bmatrix}$

$\bar{x} = (20, 24)$ $\quad$ $\bar{p} = (12, 20, 24)$

18. $\bar{a}_1 = (-3, 1, 0)$, $\bar{a}_2 = (4, 0, 1)$

$\begin{bmatrix} 10 & -12 \\ -12 & 17 \end{bmatrix} \bar{x} = \begin{bmatrix} -26 \\ 78 \end{bmatrix}$

$\bar{x} = (19, 18)$ $\quad$ $\bar{p} = (15, 19, 18)$

19. $\bar{a}_1 = (2, 1, 0, 0)$, $\bar{a}_2 = (-3, 0, 1, 0)$, $\bar{a}_3 = (4, 0, 0, 1)$

$$\begin{bmatrix} 5 & -6 & 8 \\ -6 & 10 & -12 \\ 8 & -12 & 17 \end{bmatrix} \bar{x} = \begin{bmatrix} 3 \\ -6 \\ 9 \end{bmatrix}$$

$\bar{x} = (-1,0,1)$ $\bar{p} = (2,-1,0,1)$

20. $\bar{a}_1 = (-2,1,0,0)$, $\bar{a}_2 = (-2,0,1,0)$, $\bar{a}_3 = (-2,0,0,1)$

$$\begin{bmatrix} 5 & 4 & 4 \\ 4 & 5 & 4 \\ 4 & 4 & 5 \end{bmatrix} \bar{x} = \begin{bmatrix} -13 \\ -13 \\ -13 \end{bmatrix}$$

$\bar{x} = (-1,-1,-1)$ $\bar{p} = (6,-1,-1,-1)$

21. $\bar{a}_1 = (0,-1,1,0)$, $\bar{a}_2 = (-2,1,0,1)$

$$\begin{bmatrix} 2 & -1 \\ -1 & 6 \end{bmatrix} \bar{x} = \begin{bmatrix} -11 \\ 11 \end{bmatrix}$$

$\bar{x} = (-5,1)$ $\bar{p} = (-2,6,-5,1)$

22. $\bar{a}_1 = (-2,-1,1,0)$, $\bar{a}_2 = (-1,-2,0,1)$

$$\begin{bmatrix} 6 & 4 \\ 4 & 6 \end{bmatrix} \bar{x} = \begin{bmatrix} -10 \\ -10 \end{bmatrix}$$

$\bar{x} = (-1,-1)$ $\bar{p} = (3,3,-1,-1)$

23. Here $A^T = [1 \;\; 1 \; \cdots \; 1]$, so $A^TA = m$ and $A^T\bar{b} = b_1 + b_2 + \cdots + b_m$. Hence the normal equation is

$$m\bar{x} = b_1 + b_2 + \cdots + b_m.$$

24. If $\bar{p}_1 + \bar{q}_1 = \bar{p}_2 + \bar{q}_2$ then $\bar{p}_1 - \bar{p}_2 = \bar{q}_2 - \bar{q}_1$. But $\bar{p}_1 - \bar{p}_2$ is in V and $\bar{q}_2 - \bar{q}_1$ is orthogonal to V. Hence $\bar{p}_1 - \bar{p}_2 = \bar{q}_2 - \bar{q}_1 = \bar{0}$, so $\bar{p}_1 = \bar{p}_2$ and $\bar{q}_1 = \bar{q}_2$.

25. If $\bar{b}$ is orthogonal to $\text{Col}(A) = \text{Row}(A^T)$, then $A^T\bar{b} = \bar{0}$, so the normal system is $A^TA\bar{x} = \bar{0}$. But A^TA is nonsingular by Theorem 2, so it follows that $\bar{x} = \bar{0}$.

26. We are given that $Ax^* = \bar{b}$, so Equation (13) for the orthogonal projection yields

$$A\bar{x} = \bar{p} = A(A^TA)^{-1}A^T\bar{b}$$

$$= A(A^TA)^{-1}A^TAx^* = Ax^*.$$

But the column vectors of A are linearly independent, so the fact that $A\bar{x} = Ax^*$ implies that $\bar{x} = x^*$.

27. The fact that the column vectors of A are mutually orthogonal implies that A^TA is the $n \times n$ diagonal matrix with diagonal elements $\bar{a}_1\cdot\bar{a}_1, \bar{a}_2\cdot\bar{a}_2, \cdots, \bar{a}_n\cdot\bar{a}_n$. Hence the normal equations reduce to

$$(\bar{a}_i\cdot\bar{a}_i)\bar{x}_i = \bar{a}_i\cdot\bar{b}$$

for $i = 1,2,\cdots,n$. Thus $\bar{x}_i = (\bar{a}_i\cdot\bar{b})/(\bar{a}_i\cdot\bar{a}_i)$, so the equation

$$\bar{p} = A\bar{x} = \bar{x}_1\bar{a}_1 + \bar{x}_2\bar{a}_2 + \cdots + \bar{x}_n\bar{a}_n$$

yields the desired formula for $\bar{p}$.

28. $\bar{p} = 2\bar{a}_1 + 3\bar{a}_2 + \bar{a}_3 = (6,0,4,-2)$

29. $\bar{p} = 6\bar{a}_1 + 6\bar{a}_2 + \bar{a}_3 = (7,6,5,6)$

30. In order to show that the matrices A and A^TA have the same rank, it suffices to show that they have the same null space. The reason is that

$$\text{rank} + \text{dim null space} = n$$

for each matrix. Now $A\bar{x} = \bar{0}$ implies that $A^TA\bar{x} = \bar{0}$, so the null space of A is contained in the null space of A^TA. Conversely, suppose that $A^TA\bar{x} = \bar{0}$. Then

$$0 = \bar{x}^T A^T A\bar{x} = (A\bar{x})^T A\bar{x} = (A\bar{x})\cdot(A\bar{x}) = |A\bar{x}|^2,$$

so it follows that $A\bar{x} = \bar{0}$. Thus the null space of A^TA is contained in the null space of A, and hence $\text{Null}(A) = \text{Null}(A^TA)$ as desired.

SECTION 5.3

LEAST SQUARES CURVE FITTING

This section is the reward for the somewhat heavy going in Sections 5.1 and 5.2. In the real world perhaps no applications of linear algebra are more ubiquitous than curve fitting techniques.

In Problems 1-28 we give the normal system and the equation of the indicated type that best fits the given data.

1. $$\begin{bmatrix} 3 & 6 \\ 6 & 14 \end{bmatrix}\begin{bmatrix} a \\ b \end{bmatrix} = \begin{bmatrix} 8 \\ 18 \end{bmatrix} \qquad y = \tfrac{1}{3}(2 + 3x)$$

2. $$\begin{bmatrix} 3 & 6 \\ 6 & 14 \end{bmatrix}\begin{bmatrix} a \\ b \end{bmatrix} = \begin{bmatrix} 10 \\ 23 \end{bmatrix} \qquad y = \tfrac{1}{6}(2 + 9x)$$

3. $$\begin{bmatrix} 3 & 3 \\ 3 & 11 \end{bmatrix}\begin{bmatrix} a \\ b \end{bmatrix} = \begin{bmatrix} 17 \\ 33 \end{bmatrix} \qquad y = \tfrac{1}{3}(11 + 6x)$$

4. $$\begin{bmatrix} 3 & 8 \\ 8 & 40 \end{bmatrix}\begin{bmatrix} a \\ b \end{bmatrix} = \begin{bmatrix} 12 \\ 8 \end{bmatrix} \qquad y = \tfrac{1}{7}(52 - 9x)$$

5. $\begin{bmatrix} 4 & 6 \\ 6 & 14 \end{bmatrix}\begin{bmatrix} a \\ b \end{bmatrix} = \begin{bmatrix} 21 \\ 43 \end{bmatrix}$ $\quad y = \frac{1}{10}(18 + 23x)$

6. $\begin{bmatrix} 4 & 6 \\ 6 & 14 \end{bmatrix}\begin{bmatrix} a \\ b \end{bmatrix} = \begin{bmatrix} 33 \\ 69 \end{bmatrix}$ $\quad y = \frac{3}{10}(8 + 13x)$

7. $\begin{bmatrix} 4 & 8 \\ 8 & 36 \end{bmatrix}\begin{bmatrix} a \\ b \end{bmatrix} = \begin{bmatrix} 28 \\ 96 \end{bmatrix}$ $\quad y = 3 + 2x$

8. $\begin{bmatrix} 4 & 8 \\ 8 & 36 \end{bmatrix}\begin{bmatrix} a \\ b \end{bmatrix} = \begin{bmatrix} 16 \\ -28 \end{bmatrix}$ $\quad y = 10 - 3x$

9. $\begin{bmatrix} 4 & 15 \\ 15 & 65 \end{bmatrix}\begin{bmatrix} a \\ b \end{bmatrix} = \begin{bmatrix} 31 \\ 145 \end{bmatrix}$ $\quad y = \frac{1}{7}(-32 + 23x)$

10. $\begin{bmatrix} 4 & 15 \\ 15 & 65 \end{bmatrix}\begin{bmatrix} a \\ b \end{bmatrix} = \begin{bmatrix} -45 \\ -225 \end{bmatrix}$ $\quad y = \frac{45}{7}(2 - x)$

11. $\begin{bmatrix} 4 & 6 \\ 6 & 18 \end{bmatrix}\begin{bmatrix} a \\ b \end{bmatrix} = \begin{bmatrix} 6 \\ -12 \end{bmatrix}$ $\quad y = \frac{1}{3}(15 - 7x^2)$

12. $\begin{bmatrix} 4 & 6 \\ 6 & 18 \end{bmatrix}\begin{bmatrix} a \\ b \end{bmatrix} = \begin{bmatrix} 34 \\ 80 \end{bmatrix}$ $\quad y = \frac{1}{9}(33 + 29x^2)$

13. $\begin{bmatrix} 3 & 11 \\ 11 & 83 \end{bmatrix}\begin{bmatrix} a \\ b \end{bmatrix} = \begin{bmatrix} 14 \\ 110 \end{bmatrix}$ $\quad y = \frac{1}{8}(-3 + 11x^2)$

14. $\begin{bmatrix} 3 & 14 \\ 14 & 98 \end{bmatrix}\begin{bmatrix} a \\ b \end{bmatrix} = \begin{bmatrix} 24 \\ 154 \end{bmatrix}$ $\quad y = \frac{1}{7}(14 + 9x^2)$

15. $\begin{bmatrix} 4 & 0 & 10 \\ 0 & 10 & 0 \\ 10 & 0 & 34 \end{bmatrix}\begin{bmatrix} a \\ b \\ c \end{bmatrix} = \begin{bmatrix} 18 \\ 30 \\ 54 \end{bmatrix}$ $\quad y = 2 + 3x + x^2$

16. $\begin{bmatrix} 4 & 0 & 10 \\ 0 & 10 & 0 \\ 10 & 0 & 34 \end{bmatrix}\begin{bmatrix} a \\ b \\ c \end{bmatrix} = \begin{bmatrix} 24 \\ 34 \\ 78 \end{bmatrix}$ $\quad y = \frac{1}{5}(5 + 17x + 10x^2)$

17. $\begin{bmatrix} 4 & 0 & 10 \\ 0 & 10 & 0 \\ 10 & 0 & 34 \end{bmatrix}\begin{bmatrix} a \\ b \\ c \end{bmatrix} = \begin{bmatrix} 21 \\ 30 \\ 78 \end{bmatrix}$ $\quad y = \frac{1}{6}(-11 + 18x + 17x^2)$

18. $\begin{bmatrix} 4 & 6 & 14 \\ 6 & 14 & 36 \\ 14 & 36 & 98 \end{bmatrix}\begin{bmatrix} a \\ b \\ c \end{bmatrix} = \begin{bmatrix} 18 \\ 44 \\ 124 \end{bmatrix}$ $\quad y = \frac{1}{10}(19 - 41x + 25x^2)$

19. $\begin{bmatrix} 4 & 6 & 14 \\ 6 & 14 & 36 \\ 14 & 36 & 98 \end{bmatrix}\begin{bmatrix} a \\ b \\ c \end{bmatrix} = \begin{bmatrix} 30 \\ 68 \\ 180 \end{bmatrix}$ $\quad y = \frac{1}{10}(21 + x + 15x^2)$

20. $\begin{bmatrix} 4 & 6 & 14 \\ 6 & 14 & 36 \\ 14 & 36 & 98 \end{bmatrix}\begin{bmatrix} a \\ b \\ c \end{bmatrix} = \begin{bmatrix} 0 \\ -28 \\ -100 \end{bmatrix}$ $\quad y = \frac{2}{5}(11 + 16x - 10x^2)$

21. $\begin{bmatrix} 4 & 3 & 4 \\ 3 & 5 & 2 \\ 4 & 2 & 6 \end{bmatrix}\begin{bmatrix} a \\ b \\ c \end{bmatrix} = \begin{bmatrix} 10 \\ 11 \\ 8 \end{bmatrix}$ $\quad z = \frac{1}{9}(19 + 10x - 4y)$

22. $\begin{bmatrix} 4 & 2 & 2 \\ 2 & 6 & -2 \\ 2 & -2 & 6 \end{bmatrix}\begin{bmatrix} a \\ b \\ c \end{bmatrix} = \begin{bmatrix} 10 \\ 8 \\ 0 \end{bmatrix}$ $\quad z = 3 - y$

23. $$\begin{bmatrix} 4 & 1 & 1 \\ 1 & 3 & -2 \\ 1 & -2 & 3 \end{bmatrix}\begin{bmatrix} a \\ b \\ c \end{bmatrix} = \begin{bmatrix} 14 \\ 7 \\ 0 \end{bmatrix} \qquad z = \frac{7}{10}(5 + x - y)$$

24. $$\begin{bmatrix} 4 & 3 & 3 \\ 3 & 5 & 4 \\ 3 & 4 & 5 \end{bmatrix}\begin{bmatrix} a \\ b \\ c \end{bmatrix} = \begin{bmatrix} 9 \\ 12 \\ 12 \end{bmatrix} \qquad z = \frac{1}{6}(3 + 7x + 7y)$$

25. $$\begin{bmatrix} 4 & 5 \\ 5 & 15 \end{bmatrix}\begin{bmatrix} \log a \\ k \end{bmatrix} = \begin{bmatrix} 4.6396 \\ 19.1607 \end{bmatrix} \qquad y = 0.4279\ e^{1.5270x}$$

26. $$\begin{bmatrix} 4 & 6 \\ 6 & 14 \end{bmatrix}\begin{bmatrix} \log a \\ k \end{bmatrix} = \begin{bmatrix} 3.5553 \\ -4.9618 \end{bmatrix} \qquad y = 53.37\ e^{-2.059x}$$

27. $$\begin{bmatrix} 4 & 9.6158 \\ 9.6158 & 24.2001 \end{bmatrix}\begin{bmatrix} \log a \\ k \end{bmatrix} = \begin{bmatrix} 8.2340 \\ 20.2284 \end{bmatrix}$$

$$y = 2.990\ x^{0.4006}$$

28. $$\begin{bmatrix} 4 & 3.1781 \\ 3.1781 & 3.6092 \end{bmatrix}\begin{bmatrix} \log a \\ k \end{bmatrix} = \begin{bmatrix} 2.8034 \\ 0.0317 \end{bmatrix}$$

$$y = 10.07\ x^{-2.025}$$

29. The best fit is

$$y = 207.92 - 1.25x,$$

so at a price of 69 cents per box the weekly sales should be $y(69) \approx 121.67$ thousand cases.

30. We let x denote cigarette consumption and y the cancer death rate. With both x and y in hundreds, the best fit is

$$y = 0.4075 + 0.2835x.$$

Then $y(0) = 0.4075$ and $y(4.70) = 1.7399$. Thus we would have cancer death rates of 41 per million with no cigarette consumption, and 174 per million in Australia.

31. With x and y both in hundreds, the best fit is

$$y = 21.58 + 0.7643x.$$

Thus the fixed cost is about \$2158 and the additional cost per book printed would be about 76 cents. For 2000 copies the total cost should be about \$3686.

32. The given data yield

$$\begin{bmatrix} 1 & \log 41 \\ 1 & \log 67 \\ 1 & \log 78 \\ 1 & \log 93 \\ 1 & \log 125 \end{bmatrix} \begin{bmatrix} a \\ b \end{bmatrix} = \begin{bmatrix} 89 \\ 100 \\ 103 \\ 107 \\ 110 \end{bmatrix}.$$

The best fit is $p = 17.95 + 19.38 \log w$, so $p(100) \approx 107$.

33. These normal equations result immediately when we form the normal system $A^TAx = A^Tb$ corresponding (with $k = 2$) to the linear system in (12).

34. These normal equations result immediately when we form the normal system corresponding to the linear system in (14).

SECTION 5.4

ORTHOGONAL BASES AND THE GRAM-SCHMIDT ALGORITHM

In each of Problems 1-22 we list the orthogonal basis $\{\bar{u}_i\}$ that results upon application of the Gram-Schmidt algorithm to the giveh basis $\{\bar{v}_i\}$. We have eliminated fractional factors as in Examples 3 and 4 in the text.

1. $\bar{u}_1 = (3,2)$, $\bar{u}_2 = (-2,3)$

2. $\bar{u}_1 = (3,5)$, $\bar{u}_2 = (-5,3)$

3. $\bar{u}_1 = (3,1,2)$, $\bar{u}_2 = (-11,1,16)$; $\bar{p} = (5,2,5)$

4. $\bar{u}_1 = (2,2,3)$, $\bar{u}_2 = (10,-7,-2)$; $\bar{p} = (2,4,5)$

5. $\bar{u}_1 = (2,2,2,1)$, $\bar{u}_2 = (-11,2,2,14)$; $\bar{p} = (9,12,12,9)$

6. $\bar{u}_1 = (2,-1,1,0)$, $\bar{u}_2 = (8,11,-5,6)$; $\bar{p} = (8,11,-5,6)$

7. $\bar{u}_1 = (1,1,0)$, $\bar{u}_2 = (1,-1,2)$, $\bar{u}_3 = (-1,1,1)$

8. $\bar{u}_1 = (2,1,1)$, $\bar{u}_2 = (-1,2,0)$, $\bar{u}_3 = (-2,-1,5)$

9. $\bar{u}_1 = (2,1,1)$, $\bar{u}_2 = (-4,7,1)$, $\bar{u}_3 = (-1,-1,3)$

10. $\bar{u}_1 = (1,2,2)$, $\bar{u}_2 = (10,-7,2)$, $\bar{u}_3 = (2,2,-3)$

11. $\bar{u}_1 = (1,1,0,1)$, $\bar{u}_2 = (1,-2,3,1)$, $\bar{u}_3 = (-4,3,3,1)$;
$\bar{p} = (5,-2,5,4)$

12. $\bar{u}_1 = (1,1,0,0)$, $\bar{u}_2 = (1,-1,1,0)$, $\bar{u}_3 = (1,-1,-2,3)$;
$\bar{p} = (3,7,-6,4)$

13. $\bar{u}_1 = (2,1,0,0)$, $\bar{u}_2 = (3,-6,5,0)$, $\bar{u}_3 = (1,-2,-3,7)$;
$\bar{p} = (1,-2,-3,7)$

14. $\bar{u}_1 = (1,1,1,0)$, $\bar{u}_2 = (1,0,-1,1)$, $\bar{u}_3 = (1,-2,1,0)$;
$\bar{p} = (4,3,2,1)$

15. $\bar{u}_1 = (2,1,1,0)$, $\bar{u}_2 = (4,-1,-7,6)$, $\bar{u}_3 = (7,-23,9,2)$

16. $\bar{u}_1 = (2,2,1,1)$, $\bar{u}_2 = (2,-8,11,1)$, $\bar{u}_3 = (-17,11,11,1)$

17. $\bar{u}_1 = (2,1,1,1)$, $\bar{u}_2 = (0,1,0,-1)$, $\bar{u}_3 = (-6,11,-10,11)$

18. $\bar{u}_1 = (1,-1,1,-1)$, $\bar{u}_2 = (1,3,1,-1)$, $\bar{u}_3 = (4,0,1,5)$

19. $\bar{u}_1 = (1,0,1,1,0)$, $\bar{u}_2 = (-1,3,-1,2,3)$, $\bar{u}_3 = (7,3,-1,-6,3)$

20. $\bar{u}_1 = (1,0,2,0,0)$, $\bar{u}_2 = (4,5,-2,10,0)$, $\bar{u}_3 = (-2,12,1,-5,29)$

21. $\bar{u}_1 = (1,0,0,1,0)$, $\bar{u}_2 = (-1,4,0,1,2)$, $\bar{u}_3 = (1,-4,33,-1,9)$

22. $\bar{u}_1 = (1,2,3,2,1)$, $\bar{u}_2 = (13,7,1,-12,-6)$, $\bar{u}_3 = (5,-7,2,-3,9)$

23. Reduce the coefficient matrix A to echelon form in order to find a basis $\{\bar{v}_i\}$ for the solution space. Then use the Gram-Schmidt algorithm to convert the basis $\{\bar{v}_i\}$ to an orthogonal basis $\{\bar{u}_i\}$.

24. Let A be the $k \times n$ matrix having the given mutually orthogonal vectors $\{\bar{v}_i\}$ as its row vectors. Find (as in Problem 23) an orthogonal basis $\{\bar{u}_j\}$ for the solution space of $A\bar{x} = \bar{0}$. Then the vectors $\{\bar{v}_i\}$ and $\{\bar{u}_j\}$ together form the desired orthogonal basis for R^n.

SECTION 5.5

INNER PRODUCT SPACES

The 2×2 matrices given in Problems 1-6 are symmetric, so in each case it is immediate that $\langle\bar{u},\bar{v}\rangle = \bar{u}^T A\bar{v}$ satisfies properties (i) - (iii) of an inner product. Then it follows easily from the given expression of $\bar{u}^T A\bar{u}$ as a sum of squares that the matrix is positive definite, so $\langle\bar{u},\bar{v}\rangle$ is an inner product on R^2.

1. $\bar{u}^T A\bar{u} = 2x^2 + 3y^2$

2. $\bar{u}^T A\bar{u} = (x + y)^2 + 2y^2$

3. $\bar{u}^T A\bar{u} = 2(x - y)^2 + y^2$

4. $\bar{u}^T A\bar{u} = (x + 3y)^2 + y^2$

5. $\bar{u}^T A\bar{u} = (2x + 3y)^2 + 2y^2$

6. $\bar{u}^T A\bar{u} = (3x - y)^2 + y^2$

In each of Problems 7-10 we set

$$\bar{u}_1 = \bar{e}_1 \quad \text{and} \quad \bar{u}_2 = \bar{e}_2 - \frac{\langle \bar{u}_1, \bar{e}_2 \rangle}{\langle u_1, u_1 \rangle} \bar{u}_1,$$

and then multiply $\bar{u}_2$ by a constant factor to remove any unwanted fractions.

7. $\bar{u}_1 = (1,0), \quad \bar{u}_2 = (1,1)$

8. $\bar{u}_1 = (1,0), \quad \bar{u}_2 = (-3,1)$

9. $\bar{u}_1 = (1,0), \quad \bar{u}_2 = (-3,2)$

10. $\bar{u}_1 = (1,0), \quad \bar{u}_2 = (1,3)$

Problems 11-14 are 3-dimensional analogues of Problems 1-10 above.

11. $\bar{u}^T A \bar{u} = (x + y)^2 + (x + z)^2 + y^2$
    Then $x + y = x + z = y = 0$ implies that $x = y = z = 0$.

12. $\bar{u}^T A \bar{u} = (x + 2y)^2 + (y + 2z)^2 + x^2$
    Then $x + 2y = y + 2z = x = 0$ implies that $x = y = z = 0$.

13. $\bar{u}_1 = (1,0,0), \quad \bar{u}_2 = (-1,2,0), \quad \bar{u}_3 = (-2,1,3)$

14. $\bar{u}_1 = (1,0,0), \quad \bar{u}_2 = (-1,1,0), \quad \bar{u}_3 = (2,-2,3)$

15. It is immediate that $\langle p,q \rangle$ satisfies the first three properties of an inner product. If

$$\langle p,p \rangle = p(0)p(0) + p(1)p(1) + p(2)p(2) = 0$$

then it follows that $p(0) = p(1) = p(2) = 0$. But a quadratic polynomial which vanishes at 3 distinct points must vanish identically, so $\langle p,p \rangle = 0$ implies $p = 0$. Thus $\langle p,q \rangle$ also satisfies property (iv), and therefore is an inner product.

16. With the inner product of Problem 15 we find that $\langle 1,1 \rangle = \langle x,1 \rangle = 3$, so $p_0(x) = 1$ and

$$p_1(x) = x - \frac{\langle x,1\rangle}{\langle 1,1\rangle}(1) = x - 1.$$

Then $\langle x^2,1\rangle = 5$, $\langle x^2,x - 1\rangle = 4$, and $\langle x - 1,x - 1\rangle = 2$, so

$$p_2(x) = x^2 - \frac{\langle x^2,1\rangle}{\langle 1,1\rangle}(1) - \frac{\langle x^2,x - 1\rangle}{\langle x - 1,x - 1\rangle}(x - 1)$$

$$= x^2 - 2x + (1/3).$$

17. The fact that A is positive definite if both $a > 0$ and $ac - b^2 > 0$ follows immediately from the identity

$$ax^2 + 2bxy + y^2 = a(x + \frac{b}{a}y)^2 + \frac{ac - b^2}{a}y^2.$$

18. Given

$$c_1\bar{v}_1 + c_2\bar{v}_2 + \cdots + c_n\bar{v}_n = \bar{0},$$

take the inner product of each side with $\bar{v}_k$. Then $c_k\langle\bar{v}_k,\bar{v}_k\rangle = 0$ implies that $c_k = 0$.

19. It is necessary only to replace $\bar{u}\cdot\bar{v}$ throughout with $\langle\bar{u},\bar{v}\rangle$ in the proof of Theorem 1 in Section 5.1.

20. Replace $\bar{p}$ by the righthand side of Equation (10) in $\bar{q} = \bar{b} - \bar{p}$. Then it follows immediately that $\langle\bar{q},\bar{u}_k\rangle = 0$ for each $k = 1,2,\cdots,n$. Thus $\bar{q}$ is orthogonal to each of the given basis vectors for the subspace W, and therefore is orthogonal to W.

21. First we compute

$$\langle 1,1\rangle = \int_0^1 1\,dx = 1 \quad\text{and}\quad \langle e^x,1\rangle = \int_0^1 e^x dx = e - 1.$$

Then $p_1(x) = 1$ and

$$p_2(x) = e^x - \frac{\langle e^x,1\rangle}{\langle 1,1\rangle}1 = e^x - e + 1.$$

22. Now we compute the integrals

$$\langle x, p_1(x)\rangle = \int_0^1 x\,dx = \frac{1}{2},$$

$$\langle x, p_2(x)\rangle = \int_0^1 (xe^x - ex + x)dx = \frac{1}{2}(3 - e),$$

and

$$\langle p_2, p_2\rangle = \int_0^1 (e^x - e + 1)^2 dx = -\frac{1}{2}(3 - e)(1 - e).$$

Then

$$p(x) = \frac{\langle x, p_1\rangle}{\langle p_1, p_1\rangle} p_1(x) + \frac{\langle x, p_2\rangle}{\langle p_2, p_2\rangle} p_2(x)$$

$$= \frac{1/2}{1}\cdot 1 - \frac{(3 - e)}{(3 - e)(1 - e)}(e^x - e + 1)$$

$$= -\frac{1}{2} + \frac{e^x}{e - 1}\ .$$

23. In Example 2 we computed the polynomials

$$p_0(x) = 1, \quad p_1(x) = x, \quad p_2(x) = (3x^2 - 1)/3,$$
$$\text{and} \quad p_3(x) = (5x^3 - 3x)/5,$$

and found that

$$\langle p_0, p_0\rangle = 2, \quad \langle p_1, p_1\rangle = 2/3, \quad \langle p_2, p_2\rangle = 8/45.$$

Now we compute the additional inner products

$$\langle p_0, x^4\rangle = 2/5, \qquad \langle p_1, x^4\rangle = 0$$
$$\langle p_2, x^4\rangle = 16/105, \qquad \langle p_3, x^4\rangle = 0,$$
$$\text{and} \quad \langle p_3, p_3\rangle = 8/175.$$

Then orthogonal projection of x^4 yields

$$p_4(x) = x^4 - \frac{<p_0,x^4>}{<p_0,p_0>} p_0(x) - \frac{<p_1,x^4>}{<p_1,p_1>} p_1(x)$$

$$- \frac{<p_2,x^4>}{<p_2,p_2>} p_2(x) - \frac{<p_3,x^4>}{<p_3,p_3>} p_3(x)$$

$$= x^4 - \frac{2/5}{2} (1) - \frac{0}{2/3} (1)$$

$$- \frac{16/105}{8/45} (x^2 - \tfrac{1}{3}) - \frac{0}{8/175} p_3(x)$$

$$p_4(x) = \tfrac{1}{35} (35x^4 - 30x^2 + 3).$$

24. We work initially with the polynomials of Example 2, and compute first the inner products

$$<p_0,x^3> = 0, \qquad <p_1,x^3> = 2/5,$$
$$<p_2,x^3> = 0, \qquad <p_3,x^3> = 8/175.$$

Then

$$x^3 = \frac{<p_1,x^3>}{<p_1,p_1>} p_1(x) - \frac{<p_3,x^3>}{<p_3,p_3>} p_3(x)$$

$$= \tfrac{3}{5} p_1(x) + p_3(x) = \tfrac{3}{5} P_1(x) + \tfrac{2}{5} P_3(x)$$

because $p_1(x) = P_1(x)$ and $p_3(x) = (2/5)P_3(x)$.

25. First we take $p_0(x) = 1$ and compute

$$<p_0,p_0> = \int_0^1 dx = 1, \qquad <p_0,x> = \int_0^1 x\,dx = \tfrac{1}{2}.$$

Then

$$p_1(x) = x - \frac{<p_0,x>}{<p_0,p_0>} p_0(x) = x - \tfrac{1}{2}.$$

Next we compute

$$\langle p_0, x^2\rangle = 1/3, \qquad \langle p_1, x^2\rangle = \langle p_1, p_1\rangle = 1/12.$$

Then

$$p_2(x) = x^2 - \frac{\langle p_0, x^2\rangle}{\langle p_0, p_0\rangle} p_0(x) - \frac{\langle p_1, x^2\rangle}{\langle p_1, p_1\rangle} p_1(x)$$

$$= x^2 - \frac{1/3}{1}(1) - \frac{1/12}{1/12}(x - \tfrac{1}{2}) = x^2 - x - \tfrac{1}{6}.$$

26. With the polynomials $p_0(x) = 1$ and $p_1(x) = (2x - 1)/2$ of Problem 25, we find that

$$\langle p_0, e^x\rangle = \int_0^1 e^x dx = e - 1 \quad \text{and}$$

$$\langle p_1, e^x\rangle = \int_0^1 \left[xe^x - (1/2)e^x\right] dx = \tfrac{1}{2}(3 - e).$$

Then the orthogonal projection of e^x is

$$p(x) = \frac{\langle p_0, e^x\rangle}{\langle p_0, p_0\rangle} p_0(x) + \frac{\langle p_1, e^x\rangle}{\langle p_1, p_1\rangle} p_1(x)$$

$$= \frac{e - 1}{1}(1) + \frac{(3 - e)/2}{1/12}(x - \tfrac{1}{2})$$

$$p(x) = (4e - 10) + 6(3 - e)x.$$

27. (a) The substitution $u = -x$ on $[-\pi, 0]$ yields

$$\int_{-\pi}^{\pi} g(x)dx = \int_{-\pi}^{0} g(x)dx + \int_0^{\pi} g(x)dx$$

$$= \int_0^{\pi} g(-u)du + \int_0^{\pi} g(x)dx = 2\int_0^{\pi} g(x)dx$$

because $g(-u) = g(u)$.
(b) Similar.

28. If $f(x)$ is even, then $f(x)\sin nx$ is odd, so it follows from Equation (25) and Problem 27(b) that $b_n = 0$. If

$f(x)$ is odd then $f(x)\cos nx$ is odd, so it follows from Equation (24) and Problem 27(b) that $a_n = 0$.

29. Because $f(x) = |x|$ is an even function, Problem 28(a) says that its orthogonal projection contains only cosine terms. The coefficients of these terms are

$$a_0 = \frac{1}{2\pi}\int_{-\pi}^{\pi} |x|\, dx = \frac{1}{\pi}\int_0^{\pi} x\, dx = \frac{\pi}{2}$$

and

$$a_n = \frac{1}{\pi}\int_{-\pi}^{\pi} |x| \cos nx\, dx = \frac{2}{\pi}\int_0^{\pi} x \cos nx\, dx$$

$$= \frac{2}{n^2\pi}\int_0^{n\pi} u \cos u\, du$$

$$= \frac{2}{n^2\pi}\Big[u \sin u + \cos u\Big]_0^{n\pi}$$

$$= \frac{2}{n^2\pi}\Big[(-1)^n - 1\Big],$$

because $\cos n\pi = (-1)^n$. It follows that $a_n = 0$ for $n > 0$ even, and $a_n = -4/n^2\pi$ for n odd.

31. The verification of H^{-1} is routine. We find that

$$b_0 = \langle e^x, 1\rangle = \int_0^1 e^x dx = e - 1,$$

$$b_1 = \langle e^x, x\rangle = \int_0^1 xe^x dx = 1, \quad \text{and}$$

$$b_2 = \langle e^x, x^2\rangle = \int_0^1 x^2e^x dx = e - 2.$$

Then the components of $\bar{c} = H^{-1}\bar{b}$ are

$$c_0 = 9(e - 1) - 36(1) + 30(e - 2) = 39e - 105,$$
$$c_1 = -36(e - 1) + 192(1) - 180(e - 2) = 588 - 216e,$$
$$c_2 = 30(e - 1) - 180(1) + 180(e - 2) = 210e - 570.$$

Hence the orthogonal projection of $f(x) = e^x$ into P_2 is

$$\begin{aligned} p(x) &= c_0 + c_1x + c_2x^2 \\ &= (39e - 105) + (588 - 216e)x + (210e - 570)x^2. \end{aligned}$$

CHAPTER 6

EIGENVALUES AND EIGENVECTORS

SECTION 6.1

INTRODUCTION

In each of Problems 1-26 we first find the eigenvalues of the given matrix A by solving the characteristic equation

$$|A - \lambda I| = 0.$$

All of the eigenvalues that appear in these problems are integers, so each characteristic polynomial factors readily. For each eigenvalue λ of the matrix A, we determine the associated eigenvector(s) by finding a basis for the solution space of the linear system

$$(A - \lambda I)\bar{v} = \bar{0}.$$

The notation $\lambda = * : \bar{v} = (*,*)$ below means that the eigenvalue λ corresponds to the eigenvector $\bar{v}$.

1. $\lambda_1 = 3$: $\bar{v}_1 = (2,1)$ and $\lambda_2 = 2$: $\bar{v}_2 = (1,1)$

2. $\lambda_1 = 2$: $\bar{v}_1 = (2,1)$ and $\lambda_2 = -1$: $\bar{v}_2 = (1,1)$

3. $\lambda_1 = 5$: $\bar{v}_1 = (2,1)$ and $\lambda_2 = 2$: $\bar{v}_2 = (1,1)$

4. $\lambda_1 = 2$: $\bar{v}_1 = (3,2)$ and $\lambda_2 = 1$: $\bar{v}_2 = (1,1)$

5. $\lambda_1 = 4$: $\bar{v}_1 = (3,2)$ and $\lambda_2 = 1$: $\bar{v}_2 = (1,1)$

6. $\lambda_1 = 3$: $\bar{v}_1 = (4,3)$ and $\lambda_2 = 2$: $\bar{v}_2 = (1,1)$

7. $\lambda_1 = 4$: $\bar{v}_1 = (4,3)$ and $\lambda_2 = 2$: $\bar{v}_2 = (1,1)$

8. $\lambda_1 = -1$: $\bar{v}_1 = (3,4)$ and $\lambda_2 = -2$: $\bar{v}_2 = (2,3)$

9. $\lambda_1 = 4$: $\bar{v}_1 = (5,2)$ and $\lambda_2 = 3$: $\bar{v}_2 = (2,1)$

10. $\lambda_1 = 5$: $\bar{v}_1 = (5,2)$ and $\lambda_2 = 4$: $\bar{v}_2 = (2,1)$

11. $\lambda_1 = 5$: $\bar{v}_1 = (5,7)$ and $\lambda_2 = 4$: $\bar{v}_2 = (2,3)$

12. $\lambda_1 = 4$: $\bar{v}_1 = (5,3)$ and $\lambda_2 = 3$: $\bar{v}_2 = (3,2)$

13. $\lambda_1 = 1$: $\bar{v}_1 = (0,1,-3)$; $\lambda_2 = 2$: $\bar{v}_2 = (1,0,2)$; and
$\lambda_3 = 0$: $\bar{v}_3 = (0,1,-2)$

14. $\lambda_1 = 2$: $\bar{v}_1 = (0,1,-3)$; $\lambda_2 = 5$: $\bar{v}_2 = (1,0,2)$; and
$\lambda_3 = 0$: $\bar{v}_3 = (0,1,-2)$

15. $\lambda_1 = 1$: $\bar{v}_1 = (2,1,1)$; $\lambda_2 = 2$: $\bar{v}_2 = (1,0,2)$; and
$\lambda_3 = 0$: $\bar{v}_3 = (1,1,0)$

16. $\lambda_1 = 0$: $\bar{v}_1 = (1,1,1)$; $\lambda_2 = 3$: $\bar{v}_2 = (-1,0,2)$; and
$\lambda_3 = 1$: $\bar{v}_3 = (1,1,0)$

17. $\lambda_1 = 1$: $\bar{v}_1 = (1,0,1)$; $\lambda_2 = 2$: $\bar{v}_2 = (-1,1,2)$; and
$\lambda_3 = 3$: $\bar{v}_3 = (1,0,0)$

18. $\lambda_1 = 1$: $\bar{v}_1 = (0,1,3)$; $\lambda_2 = 2$: $\bar{v}_2 = (0,1,-3)$; and
$\lambda_3 = 3$: $\bar{v}_3 = (0,2,-5)$

19. $\lambda_1 = 3$: $\bar{v}_1 = (1,0,0)$
$\lambda_2 = 1$: $\bar{v}_2 = (1,0,1)$ and $\bar{v}_3 = (-1,1,2)$
(The linearly independent eigenvectors $\bar{v}_2$ and $\bar{v}_3$ are both associated with the eigenvalue $\lambda_2 = 1$.)

20. $\lambda_1 = 2$: $\bar{v}_1 = (0,2,-5)$

$\lambda_2 = 1$: $\bar{v}_2 = (1,0,-3)$ and $\bar{v}_3 = (1,0,2)$

21. $\lambda_1 = 1$: $\bar{v}_1 = (1,1,0)$

$\lambda_2 = 2$: $\bar{v}_2 = (1,1,1)$ and $\bar{v}_3 = (-1,0,2)$

22. $\lambda_1 = 2$: $\bar{v}_1 = (1,1,1)$

$\lambda_2 = -1$: $\bar{v}_2 = (-1,0,2)$ and $\bar{v}_3 = (1,1,0)$

In Problems 23-26, note that (after a single row operation in Problem 26) the 4×4 matrix A is upper triangular. Hence $|A - \lambda I|$ is simply the product of the diagonal elements.

23. $\lambda_1 = 1$: $\bar{v}_1 = (1,0,0,0)$; $\lambda_2 = 2$: $\bar{v}_2 = (2,1,0,0)$;

$\lambda_3 = 3$: $\bar{v}_3 = (3,2,1,0)$; $\lambda_4 = 4$: $\bar{v}_4 = (4,3,2,1)$

24. $\lambda_1 = 1$: $\bar{v}_1 = (1,0,0,0)$; and $\bar{v}_2 = (1,1,0,0)$;

$\lambda_2 = 3$: $\bar{v}_3 = (2,2,1,0)$; and $\bar{v}_4 = (2,2,1,1)$

25. $\lambda_1 = 1$: $\bar{v}_1 = (1,0,0,0)$; and $\bar{v}_2 = (1,1,0,0)$;

$\lambda_2 = 2$: $\bar{v}_3 = (1,1,1,0)$; and $\bar{v}_4 = (1,1,1,1)$

26. $\lambda_1 = 1$: $\bar{v}_1 = (1,0,0,1)$; $\lambda_2 = 2$: $\bar{v}_2 = (0,1,0,0)$;

$\lambda_3 = -1$: $\bar{v}_3 = (0,0,1,0)$; $\lambda_4 = -2$: $\bar{v}_4 = (1,0,0,2)$

27. If $A\bar{v} = \lambda\bar{v}$ and we assume inductively that $A^{n-1}\bar{v} = \lambda^{n-1}\bar{v}$, then multiplication by A yields

$$A^n\bar{v} = \lambda^{n-1}(A\bar{v}) = \lambda^{n-1}(\lambda\bar{v}) = \lambda^n\bar{v}.$$

Thus λ^n is an eigenvalue of the matrix A^n corresponding to the eigenvector $\bar{v}$.

28. By the remark following Example 6, any eigenvalue of an invertible matrix is nonzero. If $\lambda \neq 0$ is an eigenvalue of the invertible matrix A with associated eigenvector $\bar{v}$, then

$$\begin{aligned} A\bar{v} &= \lambda\bar{v} \\ \bar{v} &= \lambda A^{-1}\bar{v} \\ A^{-1}\bar{v} &= \lambda^{-1}\bar{v}. \end{aligned}$$

Thus λ^{-1} is an eigenvalue of A^{-1} with associated eigenvector $\bar{v}$.

29. (a) Note first that $(A - \lambda I)^T = (A^T - \lambda I)$. Because the determinant of a square matrix equals the determinant of its transpose, it follows that $|A - \lambda I| = |A^T - \lambda I|$. Thus the matrices A and A^T have the same characteristic equation, and hence have the same eigenvalues.

(b) Consider the matrix

$$A = \begin{bmatrix} 1 & 0 \\ 1 & 1 \end{bmatrix} \quad \text{with} \quad A^T = \begin{bmatrix} 1 & 1 \\ 0 & 1 \end{bmatrix}.$$

Then the only eigenvector of A is $(0,1)$, whereas the only eigenvector of A^T is $(1,0)$.

30. If the $n \times n$ matrix $A = [a_{ij}]$ is either upper or lower triangular, then obviously its characteristic equation is

$$(a_{11} - \lambda)(a_{22} - \lambda) \cdots (a_{nn} - \lambda) = 0.$$

Hence its eigenvalues are the diagonal elements $a_{11}, a_{22}, \ldots, a_{nn}$.

31. If

$$|A - \lambda I| = (-1)^n\lambda^n + c_{n-1}\lambda^{n-1} + \cdots + c_1\lambda + c_0,$$

then substitution of $\lambda = 0$ yields

$$c_0 = |A - 0I| = |A|.$$

32. The characteristic equation of the 2×2 matrix

$$A = \begin{bmatrix} a & b \\ c & d \end{bmatrix}$$

is

$$(a - \lambda)(d - \lambda) - bc = 0,$$

that is,

$$\lambda^2 - (a + d)\lambda + (ad - bc) = 0.$$

Thus the coefficient of λ is $-(a + d) = -(\text{tr } A)$.

33. If the characteristic equation is written in factored form,

$$(\lambda - \lambda_1)(\lambda - \lambda_2) \cdots (\lambda - \lambda_n) = 0,$$

then upon multiplying the factors it is clear that the coefficient of λ^{n-1} is $-(\lambda_1 + \lambda_2 + \cdots + \lambda_n)$. But according to the result stated in Problem 32, this coefficient also equals $-(\text{tr } A)$. Therefore

$$\lambda_1 + \lambda_2 + \cdots + \lambda_n = a_{11} + a_{22} + \cdots + a_{nn}.$$

34. We find that $\text{tr } A = 12$ and $\det A = 60$, so

$$p(\lambda) = -\lambda^3 + 12\lambda^2 + c_1\lambda + 60.$$

Substitution of $\lambda = 1$ and $p(1) = |A - I| = 24$ yields $c_1 = -47$, so the characteristic equation is

$$\lambda^3 - 12\lambda^2 + 47\lambda - 60 = 0.$$

Trying $\lambda = \pm 1, \pm 2, \pm 3$ in turn, we discover the eigenvalue $\lambda_1 = 3$. Then division of the cubic by $(\lambda - 3)$ yields

$$\lambda^2 - 9\lambda + 20 = (\lambda - 4)(\lambda - 5),$$

so the other two eigenvalues are $\lambda_2 = 4$ and $\lambda_3 = 5$.

Finally, we find that the eigenvectors associated with the eigenvalues λ_1, λ_2, and λ_3 are $\bar{v}_1 = (3,2,1)$, $\bar{v}_2 = (5,7,7)$, and $\bar{v}_3 = (-1,1,2)$, respectively.

35. We find that $\operatorname{tr} A = 8$ and $\det A = -60$, so

$$p(\lambda) = \lambda^4 - 8\lambda^3 + c_2\lambda^2 + c_1\lambda - 60.$$

Substitution of $\lambda = 1$, $p(1) = |A - I| = -24$ and $\lambda = -1$, $p(-1) = |A + I| = -72$ yields the equations

$$\begin{aligned} c_2 + c_1 &= 43 \\ c_2 - c_1 &= -21, \end{aligned}$$

which we solve for $c_1 = 32$ and $c_2 = 11$. Hence the characteristic equation of A is

$$\lambda^4 - 8\lambda^3 + 11\lambda^2 + 32\lambda - 60 = 0.$$

Trying $\lambda = \pm 1, \pm 2$ in turn, we discover the eigenvalues $\lambda_1 = 2$ and $\lambda_2 = -2$. Then division of the 4th degree polynomial by $(\lambda^2 - 4)$ yields

$$\lambda^2 - 8\lambda + 15 = (\lambda - 3)(\lambda - 5),$$

so the other two eigenvalues are $\lambda_3 = 3$ and $\lambda_4 = 5$. Finally we find that the eigenvectors associated with these four eigenvalues are as follows --

$$\begin{aligned} &\lambda_1 = 2: \ \bar{v}_1 = (3,4,0,3) \qquad &&\lambda_2 = -2: \ \bar{v}_2 = (1,0,1,2) \\ &\lambda_3 = 3: \ \bar{v}_3 = (3,1,2,4) \qquad &&\lambda_4 = 5: \ \bar{v}_4 = (1,1,0,1) \end{aligned}$$

SECTION 6.2

DIAGONALIZATION OF MATRICES

In each of Problems 1-28 where the given matrix A is diagonalizable, we list the diagonalizing matrix P and the diagonal matrix D such that $P^{-1}AP = D$. In this case the column vectors of P are the eigenvectors of A and the diagonal elements of D are the corresponding eigenvalues. In those problems where the given matrix A is not diagonalizable, we briefly explain why not.

1. $P = \begin{bmatrix} 2 & 1 \\ 1 & 1 \end{bmatrix}$ $D = \begin{bmatrix} 3 & 0 \\ 0 & 1 \end{bmatrix}$

2. $P = \begin{bmatrix} 3 & 1 \\ 2 & 1 \end{bmatrix}$ $D = \begin{bmatrix} 2 & 0 \\ 0 & 0 \end{bmatrix}$

3. $P = \begin{bmatrix} 3 & 1 \\ 2 & 1 \end{bmatrix}$ $D = \begin{bmatrix} 3 & 0 \\ 0 & 2 \end{bmatrix}$

4. $P = \begin{bmatrix} 4 & 1 \\ 3 & 1 \end{bmatrix}$ $D = \begin{bmatrix} 2 & 0 \\ 0 & 1 \end{bmatrix}$

5. $P = \begin{bmatrix} 4 & 1 \\ 3 & 1 \end{bmatrix}$ $D = \begin{bmatrix} 3 & 0 \\ 0 & 1 \end{bmatrix}$

6. $P = \begin{bmatrix} 3 & 2 \\ 4 & 3 \end{bmatrix}$ $D = \begin{bmatrix} 2 & 0 \\ 0 & 1 \end{bmatrix}$

7. $P = \begin{bmatrix} 5 & 2 \\ 2 & 1 \end{bmatrix}$ $D = \begin{bmatrix} 2 & 0 \\ 0 & 1 \end{bmatrix}$

8. $P = \begin{bmatrix} 5 & 3 \\ 3 & 2 \end{bmatrix}$ $D = \begin{bmatrix} 2 & 0 \\ 0 & 1 \end{bmatrix}$

9. The given matrix A has only one eigenvalue $\lambda = 1$ and only one associated eigenvector $\bar{v} = (2,1)$, and therefore is not diagonalizable.

10. The given matrix A has only one eigenvalue $\lambda = 2$ and only

one associated eigenvector $\bar{v} = (1,1)$, and therefore is not diagonalizable.

11. The given matrix A has only one eigenvalue $\lambda = 2$ and only one associated eigenvector $\bar{v} = (1,-3)$, and therefore is not diagonalizable.

12. The given matrix A has only one eigenvalue $\lambda = -1$ and only one associated eigenvector $\bar{v} = (3,-4)$, and therefore is not diagonalizable.

13. $P = \begin{bmatrix} 0 & 3 & 1 \\ 0 & 1 & 0 \\ 1 & 2 & 0 \end{bmatrix}$ $\quad D = \begin{bmatrix} 2 & 0 & 0 \\ 0 & 2 & 0 \\ 0 & 0 & 1 \end{bmatrix}$

14. $P = \begin{bmatrix} 1 & -1 & 1 \\ 1 & 0 & 1 \\ 1 & 2 & 0 \end{bmatrix}$ $\quad D = \begin{bmatrix} 1 & 0 & 0 \\ 0 & 0 & 0 \\ 0 & 0 & 0 \end{bmatrix}$

15. $P = \begin{bmatrix} 1 & -1 & 1 \\ 1 & 0 & 1 \\ 1 & 2 & 0 \end{bmatrix}$ $\quad D = \begin{bmatrix} 1 & 0 & 0 \\ 0 & 1 & 0 \\ 0 & 0 & 0 \end{bmatrix}$

16. $P = \begin{bmatrix} 1 & -1 & 1 \\ 1 & 0 & 1 \\ 1 & 2 & 0 \end{bmatrix}$ $\quad D = \begin{bmatrix} 1 & 0 & 0 \\ 0 & 3 & 0 \\ 0 & 0 & 1 \end{bmatrix}$

17. $P = \begin{bmatrix} 1 & -1 & 1 \\ 1 & 0 & 1 \\ 1 & 2 & 0 \end{bmatrix}$ $\quad D = \begin{bmatrix} 2 & 0 & 0 \\ 0 & 1 & 0 \\ 0 & 0 & -1 \end{bmatrix}$

18. $P = \begin{bmatrix} 1 & -1 & 1 \\ 1 & 0 & 1 \\ 1 & 2 & 0 \end{bmatrix}$ $\quad D = \begin{bmatrix} 3 & 0 & 0 \\ 0 & 2 & 0 \\ 0 & 0 & 1 \end{bmatrix}$

19. $P = \begin{bmatrix} 1 & -1 & 1 \\ 1 & 0 & 1 \\ 1 & 2 & 0 \end{bmatrix}$ $\quad D = \begin{bmatrix} 1 & 0 & 0 \\ 0 & 3 & 0 \\ 0 & 0 & 2 \end{bmatrix}$

20. $P = \begin{bmatrix} 1 & 0 & 0 \\ 0 & 1 & 2 \\ 3 & -3 & -5 \end{bmatrix}$ $\qquad D = \begin{bmatrix} 2 & 0 & 0 \\ 0 & 5 & 0 \\ 0 & 0 & 6 \end{bmatrix}$

21. The given matrix A has the single eigenvalue $\lambda = 1$ and only two associated eigenvectors, $\bar{v}_1 = (1,1,0)$ and $\bar{v}_2 = (0,0,1)$, and therefore is not diagonalizable.

22. The given matrix A has the single eigenvalue $\lambda = 1$ and the single eigenvector $\bar{v} = (1,1,1)$, and therefore is not diagonalizable.

23. The given matrix A has the eigenvalue $\lambda_1 = 1$ with the single eigenvector $\bar{v}_1 = (1,1,1)$, and the eigenvalue $\lambda_2 = 2$ with the single eigenvector $\bar{v}_2 = (1,1,0)$. Thus A has only two linearly independent eigenvectors, and therefore is not diagonalizable.

24. The given matrix A has the eigenvalue $\lambda_1 = 1$ with the single eigenvector $\bar{v}_1 = (1,1,0)$, and the eigenvalue $\lambda_2 = 2$ with the single eigenvector $\bar{v}_2 = (1,1,1)$. Therefore A is not diagonalizable.

25. $P = \begin{bmatrix} 1 & 1 & 1 & 1 \\ 0 & 1 & 1 & 1 \\ 0 & 0 & 1 & 1 \\ 0 & 0 & 0 & 1 \end{bmatrix}$ $\qquad D = \begin{bmatrix} 1 & 0 & 0 & 0 \\ 0 & 1 & 0 & 0 \\ 0 & 0 & -1 & 0 \\ 0 & 0 & 0 & -1 \end{bmatrix}$

26. $P = \begin{bmatrix} 1 & 1 & 1 & 1 \\ 0 & 1 & 1 & 1 \\ 0 & 0 & 1 & 1 \\ 0 & 0 & 0 & 1 \end{bmatrix}$ $\qquad D = \begin{bmatrix} 1 & 0 & 0 & 0 \\ 0 & 1 & 0 & 0 \\ 0 & 0 & 1 & 0 \\ 0 & 0 & 0 & 2 \end{bmatrix}$

27. The given matrix A has the eigenvalue $\lambda_1 = 1$ with the single eigenvector $\bar{v}_1 = (1,0,0,0)$, and the eigenvalue $\lambda_2 = 2$ with the single eigenvector $\bar{v}_2 = (1,1,1,1)$. Thus the 4×4 matrix A has only two linearly independent eigenvectors, and therefore is not diagonalizable.

28. The given matrix A has the eigenvalue $\lambda_1 = 1$ with the single eigenvector $\bar{v}_1 = (1,0,0,0)$, and the eigenvalue $\lambda_2 = 2$ with the single eigenvector $\bar{v}_2 = (1,1,1,0)$. Therefore A is not diagonalizable.

29. If

$$A = P^{-1}BP \quad \text{and} \quad B = Q^{-1}CQ$$

then

$$A = P^{-1}(Q^{-1}CQ)P = (QP)^{-1}C(QP).$$

30. If $A = P^{-1}BP$ then

$$A^n = (P^{-1}BP)(P^{-1}BP)(P^{-1}BP)\cdots(P^{-1}BP)(P^{-1}BP)$$

$$= P^{-1}BIBI\cdots BIBP = P^{-1}B^nP.$$

31. If $A = P^{-1}BP$ then $A^{-1} = (P^{-1}BP)^{-1} = P^{-1}B^{-1}P$.

32. If $A = P^{-1}BP$ then $|P^{-1}||P| = 1$ so

$$|A - \lambda I| = |P^{-1}||A - \lambda I||P|$$

$$= |P^{-1}(A - \lambda I)P|$$

$$= |P^{-1}AP - \lambda P^{-1}IP| = |B - \lambda I|.$$

Thus A and B have the same characteristic polynomial.

33. If $A = P^{-1}BP$ then $|A| = |P|^{-1}|B||P| = |B|$. Moreover, A and B have the same eigenvalues $\lambda_1,\lambda_2,\cdots,\lambda_n$, so

$$\text{tr } A = a_{11} + a_{22} + \cdots + a_{nn}$$

$$= \lambda_1 + \lambda_2 + \cdots + \lambda_n$$

$$= b_{11} + b_{22} + \cdots + b_{nn} = \text{tr } B.$$

34. The characteristic equation of A is

$$\lambda^2 - (a + d)\lambda + (ad - bc) = 0,$$

and the discriminant of this quadratic equation is

$$\Delta = (a + d)^2 - 4(ad - bc) = (a - d)^2 + 4bc.$$

If $\Delta > 0$ then A has two distinct eigenvalues and hence two linearly independent eigenvectors, and is therefore diagonalizable. If $\Delta < 0$ then A has no (real) eigenvalues and no eigenvectors, and therefore is not diagonalizable. Finally, note that $\Delta = 0$ for both

$$I = \begin{bmatrix} 1 & 0 \\ 0 & 1 \end{bmatrix} \quad \text{and} \quad A = \begin{bmatrix} 1 & 1 \\ 0 & 1 \end{bmatrix},$$

but A has only the single eigenvalue $\lambda = 1$ and single associated eigenvector $\bar{v} = (1,0)$, and therefore is not diagonalizable.

35. Three eigenvectors associated with the three distinct eigenvalues can be arranged in six different orders as the column vectors of the diagonalizing matrix $P = [\bar{v}_1 \;\; \bar{v}_2 \;\; \bar{v}_3]$.

36. The fact that the matrices A and B have the same eigenvalues implies that they are similar to the same diagonal matrix D having these eigenvalues as its diagonal elements. But two matrices that are similar to a third matrix are (by Problem 29) similar to each other.

37. This follows immediately from the fact that if $A = PDP^{-1}$ then

$$A^2 = (PDP^{-1})(PDP^{-1}) = PD^2P^{-1}.$$

Thus the same (eigenvector) matrix P diagonalizes A^2, but the resulting diagonal (eigenvalue) matrix is the square of the one for A.

38. Under the stated assumption, $A = PDP^{-1}$ with $D = \lambda I$, so

$$A = P(\lambda I)P^{-1} = \lambda PIP^{-1} = \lambda I = D.$$

39. Let A have k distinct eigenvalues $\lambda_1, \lambda_2, \cdots, \lambda_k$. Then the definition of algebraic multiplicity and the fact that all solutions of the nth degree polynomial equation $|A - \lambda I| = 0$ are real imply that

$$p_1 + p_2 + \cdots + p_k = n.$$

Now Theorem 4 implies that A is diagonalizable if and only if

$$q_1 + q_2 + \cdots + q_k = n.$$

But, because $p_i \geq q_i$ for each i, this is so if and only if $p_i = q_i$ for each $i = 1, 2, \cdots, k$.

SECTION 6.3

APPLICATIONS INVOLVING POWERS OF MATRICES

In each of Problems 1-14 we list the matrices P and D, and the desired power $A^k = PD^kP^{-1}$.

1. $P = \begin{bmatrix} 2 & 1 \\ 1 & 1 \end{bmatrix}$ $D = \begin{bmatrix} 2 & 0 \\ 0 & 1 \end{bmatrix}$ $A^5 = \begin{bmatrix} 63 & -62 \\ 31 & -30 \end{bmatrix}$

2. $P = \begin{bmatrix} 2 & 1 \\ 1 & 1 \end{bmatrix}$ $D = \begin{bmatrix} 2 & 0 \\ 0 & -1 \end{bmatrix}$ $A^5 = \begin{bmatrix} 65 & -66 \\ 33 & -34 \end{bmatrix}$

3. $P = \begin{bmatrix} 3 & 1 \\ 2 & 1 \end{bmatrix}$ $D = \begin{bmatrix} 2 & 0 \\ 0 & 0 \end{bmatrix}$ $A^5 = \begin{bmatrix} 96 & -96 \\ 64 & -64 \end{bmatrix}$

4. $P = \begin{bmatrix} 3 & 1 \\ 2 & 1 \end{bmatrix}$ $D = \begin{bmatrix} 2 & 0 \\ 0 & 1 \end{bmatrix}$ $A^5 = \begin{bmatrix} 94 & -93 \\ 62 & -61 \end{bmatrix}$

5. $P = \begin{bmatrix} 4 & 1 \\ 3 & 1 \end{bmatrix}$ $D = \begin{bmatrix} 2 & 0 \\ 0 & 1 \end{bmatrix}$ $A^5 = \begin{bmatrix} 125 & -124 \\ 93 & -92 \end{bmatrix}$

6. $P = \begin{bmatrix} 5 & 2 \\ 2 & 1 \end{bmatrix}$ $D = \begin{bmatrix} 2 & 0 \\ 0 & 1 \end{bmatrix}$ $A^5 = \begin{bmatrix} 156 & -310 \\ 62 & -123 \end{bmatrix}$

7. $P = \begin{bmatrix} 0 & 3 & 1 \\ 0 & 1 & 0 \\ 1 & 2 & 0 \end{bmatrix}$ $D = \begin{bmatrix} 2 & 0 & 0 \\ 0 & 2 & 0 \\ 0 & 0 & 1 \end{bmatrix}$ $A^5 = \begin{bmatrix} 1 & 93 & 0 \\ 0 & 32 & 0 \\ 0 & 0 & 32 \end{bmatrix}$

8. $P = \begin{bmatrix} 1 & -1 & 1 \\ 0 & 1 & 0 \\ 1 & 2 & 0 \end{bmatrix}$ $D = \begin{bmatrix} 2 & 0 & 0 \\ 0 & 1 & 0 \\ 0 & 0 & 1 \end{bmatrix}$ $A^5 = \begin{bmatrix} 1 & -62 & 31 \\ 0 & 1 & 0 \\ 0 & -62 & 32 \end{bmatrix}$

9. $P = \begin{bmatrix} 1 & -1 & 1 \\ 0 & 1 & 0 \\ 1 & 2 & 0 \end{bmatrix}$ $D = \begin{bmatrix} 2 & 0 & 0 \\ 0 & 2 & 0 \\ 0 & 0 & 1 \end{bmatrix}$ $A^5 = \begin{bmatrix} 1 & -93 & 31 \\ 0 & 32 & 0 \\ 0 & 0 & 32 \end{bmatrix}$

10. $P = \begin{bmatrix} 0 & -1 & 1 \\ 1 & 0 & 1 \\ 1 & 2 & 0 \end{bmatrix}$ $D = \begin{bmatrix} 2 & 0 & 0 \\ 0 & 2 & 0 \\ 0 & 0 & 1 \end{bmatrix}$ $A^5 = \begin{bmatrix} 94 & -93 & 31 \\ 62 & -61 & 31 \\ 0 & 0 & 32 \end{bmatrix}$

11. $P = \begin{bmatrix} 1 & 0 & 0 \\ 0 & 1 & 2 \\ 3 & -3 & -5 \end{bmatrix}$ $D = \begin{bmatrix} 1 & 0 & 0 \\ 0 & -1 & 0 \\ 0 & 0 & 0 \end{bmatrix}$ $A^{10} = \begin{bmatrix} 1 & 0 & 0 \\ 6 & -5 & -2 \\ -15 & 15 & 6 \end{bmatrix}$

12. $P = \begin{bmatrix} 2 & 1 & 1 \\ 3 & 1 & 2 \\ 1 & 2 & 0 \end{bmatrix}$ $D = \begin{bmatrix} 1 & 0 & 0 \\ 0 & 1 & 0 \\ 0 & 0 & -1 \end{bmatrix}$ $A^{10} = \begin{bmatrix} 1 & 0 & 0 \\ 0 & 1 & 0 \\ 0 & 0 & 1 \end{bmatrix}$

13. $P = \begin{bmatrix} 1 & -1 & 1 \\ 1 & 0 & 1 \\ 1 & 2 & 0 \end{bmatrix}$ $D = \begin{bmatrix} 1 & 0 & 0 \\ 0 & -1 & 0 \\ 0 & 0 & 0 \end{bmatrix}$ $A^{10} = \begin{bmatrix} 3 & -3 & 1 \\ 2 & -2 & 1 \\ 0 & 0 & 1 \end{bmatrix}$

14. $P = \begin{bmatrix} 2 & 1 & 1 \\ 1 & 0 & 1 \\ 1 & 2 & 0 \end{bmatrix}$ $D = \begin{bmatrix} 1 & 0 & 0 \\ 0 & -1 & 0 \\ 0 & 0 & 0 \end{bmatrix}$ $A^{10} = \begin{bmatrix} 3 & -3 & -1 \\ 2 & -2 & -1 \\ 0 & 0 & 1 \end{bmatrix}$

The computations in each of Problems 15-20 follow those of Example 2 in the text. In each case the power $A^k = PD^kP^{-1}$ approaches as $k \longrightarrow \infty$ a matrix of the form

$$\begin{bmatrix} p & p \\ q & q \end{bmatrix}$$

where $p + q = 1$, p is the fraction of the population in the city, and q is the fraction in the suburbs in the long term. For each problem we give the matrices P and D and the long term distribution of population between the city and suburbs.

15. $P = \begin{bmatrix} 1 & -1 \\ 1 & 1 \end{bmatrix}$ $D = \begin{bmatrix} 1 & 0 \\ 0 & 0.8 \end{bmatrix}$ 50% city, 50% suburbs

16. $P = \begin{bmatrix} 1 & -1 \\ 3 & 1 \end{bmatrix}$ $D = \begin{bmatrix} 1 & 0 \\ 0 & 0.8 \end{bmatrix}$ 25% city, 75% suburbs

17. $P = \begin{bmatrix} 3 & -1 \\ 5 & 1 \end{bmatrix}$ $D = \begin{bmatrix} 1 & 0 \\ 0 & 0.6 \end{bmatrix}$ 3/8 city, 5/8 suburbs

18. $P = \begin{bmatrix} 1 & -1 \\ 2 & 1 \end{bmatrix}$ $D = \begin{bmatrix} 1 & 0 \\ 0 & 0.7 \end{bmatrix}$ 1/3 city, 2/3 suburbs

19. $P = \begin{bmatrix} 1 & -1 \\ 2 & 1 \end{bmatrix}$ $D = \begin{bmatrix} 1 & 0 \\ 0 & 0.85 \end{bmatrix}$ 1/3 city, 2/3 suburbs

20. $P = \begin{bmatrix} 3 & -1 \\ 4 & 1 \end{bmatrix}$ $D = \begin{bmatrix} 1 & 0 \\ 0 & 0.65 \end{bmatrix}$ 3/7 city, 4/7 suburbs

21. With $r = 0.16$ we find that

$$P = \begin{bmatrix} 5 & 5 \\ 4 & 2 \end{bmatrix} \qquad D = \begin{bmatrix} 1 & 0 \\ 0 & 0.8 \end{bmatrix}.$$

Hence when n is sufficiently large it follows that

$$A^n \approx \begin{bmatrix} 5 & 5 \\ 4 & 2 \end{bmatrix} \begin{bmatrix} 1 & 0 \\ 0 & 0 \end{bmatrix} \left(-\frac{1}{10}\right) \begin{bmatrix} 2 & -5 \\ -4 & 5 \end{bmatrix} = \frac{1}{10} \begin{bmatrix} -10 & 25 \\ -8 & 20 \end{bmatrix}.$$

Therefore as $n \longrightarrow \infty$ the fox-rabbit population approaches a stable situation with $2.5R_0 - F_0$ foxes and $2R_0 - 0.8F_0$ rabbits.

22. With $r = 0.175$ we find that

$$P = \begin{bmatrix} 10 & 2 \\ 7 & 1 \end{bmatrix} \qquad D = \begin{bmatrix} 0.95 & 0 \\ 0 & 0.85 \end{bmatrix}.$$

Hence when n is sufficiently large it follows that

$$A^n \approx \begin{bmatrix} 10 & 2 \\ 7 & 1 \end{bmatrix} \begin{bmatrix} 0 & 0 \\ 0 & 0 \end{bmatrix} \left(-\frac{1}{4}\right) \begin{bmatrix} 1 & -2 \\ -7 & 10 \end{bmatrix} = \begin{bmatrix} 0 & 0 \\ 0 & 0 \end{bmatrix}.$$

Thus the fox and rabbit populations both die out as $n \longrightarrow \infty$.

23. With $r = 0.135$ we find that

$$P = \begin{bmatrix} 10 & 10 \\ 9 & 3 \end{bmatrix} \qquad D = \begin{bmatrix} 1.05 & 0 \\ 0 & 0.75 \end{bmatrix}.$$

Hence when n is sufficiently large it follows that

$$A^n \approx -\frac{1}{60} \begin{bmatrix} 10 & 10 \\ 9 & 3 \end{bmatrix} \begin{bmatrix} (1.05)^n & 0 \\ -9 & 10 \end{bmatrix} \begin{bmatrix} 3 & -10 \\ -9 & 10 \end{bmatrix}.$$

$$= \frac{1}{60} (1.05)^n \begin{bmatrix} -30 & 100 \\ -27 & 90 \end{bmatrix}.$$

It follows that if n is large then the fox and rabbit populations both increase by 5% each month, maintaining a constant ratio of 10 foxes per 9 rabbits.

24. $A = PDP^{-1} = \begin{bmatrix} 3 & 5 \\ 4 & 7 \end{bmatrix}\begin{bmatrix} 1 & 0 \\ 0 & -1 \end{bmatrix}\begin{bmatrix} 7 & -5 \\ -4 & 3 \end{bmatrix} = \begin{bmatrix} 41 & -30 \\ 56 & -41 \end{bmatrix}$.

If n is even than $D^n = I$ so

$$A^n = PD^nP^{-1} = PIP^{-1} = PP^{-1} = I.$$

If n is odd then $A^n = A^{n-1}A = IA = A$. Thus $A^{99} = A$ and $A^{100} = I$.

25. The fact that each $|\lambda| = 1$ implies that $D^n = I$ if n is even, in which case $A^n = PD^nP^{-1} = PP^{-1} = I$.

26. We find immediately that $A^2 = I$, so $A^3 = A^2A = A$, $A^4 = A^3A = A^2 = I$, and so forth.

27. We find immediately that $A^2 = -I$, so $A^3 = A^2A = -A$, $A^4 = A^3A = -A^2 = I$, and so forth.

28. If

$$A = \begin{bmatrix} 1 & 1 \\ 0 & 1 \end{bmatrix} = \begin{bmatrix} 1 & 0 \\ 0 & 1 \end{bmatrix} + \begin{bmatrix} 0 & 1 \\ 0 & 0 \end{bmatrix} = I + B$$

then $B^2 = 0$, so it follows that

$$\begin{aligned} A^n = (I + B)^n &= I^n + nI^{n-1}B + \cdots \\ &= I + nB \\ &= \begin{bmatrix} 1 & n \\ 0 & 1 \end{bmatrix}. \end{aligned}$$

29. The characteristic equation of A is

$$\begin{aligned} &(p - \lambda)(q - \lambda) - (1 - p)(1 - q) \\ &\qquad = \lambda^2 - (p + q)\lambda + (p + q - 1) \end{aligned}$$

$$= (\lambda - 1)[\lambda - (p + q - 1)] = 0$$

so the eigenvalues are $\lambda_1 = 1$ and $\lambda_2 = p + q - 1$.

30. The fact that the column sums of A are each 1 implies that the row sums of the transpose A^T are each 1, so it follows readily that $A^T\bar{v} = \bar{v}$. Thus $\lambda = 1$ is an eigenvalue of A^T. But A and A^T have the same eigenvalues (by Problem 29 in Section 6.1).

SECTION 6.4

SYMMETRIC MATRICES AND ORTHOGONAL EIGENVECTORS

In each of Problems 1-18 we list a matrix P with orthogonal columns and a diagonal matrix D such that $PDP^{-1} = A$. The column vectors of P form a complete set of orthogonal eigenvectors of the given matrix A, and the diagonal elements of D are the corresponding eigenvalues of A. If each column vector of P is normalized, the result is an orthogonal matrix Q that diagonalizes A, that is, $QDQ^T = A$ and $Q^TAQ = D$.

1. $P = \begin{bmatrix} 1 & 1 \\ 1 & -1 \end{bmatrix}$ $D = \begin{bmatrix} 4 & 0 \\ 0 & 2 \end{bmatrix}$

2. $P = \begin{bmatrix} 1 & 1 \\ 1 & -1 \end{bmatrix}$ $D = \begin{bmatrix} 6 & 0 \\ 0 & 2 \end{bmatrix}$

3. $P = \begin{bmatrix} 2 & -1 \\ 1 & 2 \end{bmatrix}$ $D = \begin{bmatrix} 10 & 0 \\ 0 & 5 \end{bmatrix}$

4. $P = \begin{bmatrix} 2 & -1 \\ 1 & 2 \end{bmatrix}$ $D = \begin{bmatrix} 15 & 0 \\ 0 & 5 \end{bmatrix}$

5. $P = \begin{bmatrix} 1 & 0 & 1 \\ 1 & 0 & -1 \\ 0 & 1 & 0 \end{bmatrix}$ $D = \begin{bmatrix} 4 & 0 & 0 \\ 0 & 2 & 0 \\ 0 & 0 & -2 \end{bmatrix}$

6. $P = \begin{bmatrix} 1 & 0 & 1 \\ 1 & 0 & -1 \\ 0 & 1 & 0 \end{bmatrix}$ $D = \begin{bmatrix} 8 & 0 & 0 \\ 0 & 4 & 0 \\ 0 & 0 & 2 \end{bmatrix}$

7. $P = \begin{bmatrix} 1 & 1 & 1 \\ 1 & -2 & 0 \\ 1 & 1 & -1 \end{bmatrix}$ $D = \begin{bmatrix} 6 & 0 & 0 \\ 0 & 6 & 0 \\ 0 & 0 & 0 \end{bmatrix}$

8. $P = \begin{bmatrix} 1 & 1 & 1 \\ -1 & 1 & 1 \\ 0 & -2 & 1 \end{bmatrix}$ $D = \begin{bmatrix} 9 & 0 & 0 \\ 0 & 3 & 0 \\ 0 & 0 & 3 \end{bmatrix}$

9. $P = \begin{bmatrix} 1 & 1 & -1 \\ 0 & 1 & 2 \\ 1 & -1 & 1 \end{bmatrix}$ $D = \begin{bmatrix} 6 & 0 & 0 \\ 0 & 3 & 0 \\ 0 & 0 & 0 \end{bmatrix}$

10. $P = \begin{bmatrix} 1 & 1 & -1 \\ 0 & 1 & 2 \\ 1 & -1 & 1 \end{bmatrix}$ $D = \begin{bmatrix} 6 & 0 & 0 \\ 0 & 6 & 0 \\ 0 & 0 & 0 \end{bmatrix}$

11. $P = \begin{bmatrix} 1 & 1 & 1 \\ 1 & -2 & 0 \\ 1 & 1 & -1 \end{bmatrix}$ $D = \begin{bmatrix} 9 & 0 & 0 \\ 0 & 6 & 0 \\ 0 & 0 & 0 \end{bmatrix}$

12. $P = \begin{bmatrix} 1 & 0 & 4 \\ 2 & 1 & -1 \\ 2 & -1 & -1 \end{bmatrix}$ $D = \begin{bmatrix} 9 & 0 & 0 \\ 0 & 6 & 0 \\ 0 & 0 & 0 \end{bmatrix}$

13. $P = \begin{bmatrix} 1 & 1 & 1 \\ 1 & -2 & 0 \\ 1 & 1 & -1 \end{bmatrix}$ $D = \begin{bmatrix} 6 & 0 & 0 \\ 0 & 3 & 0 \\ 0 & 0 & 3 \end{bmatrix}$

14. $P = \begin{bmatrix} 1 & 1 & 1 \\ 1 & -2 & 0 \\ 1 & 1 & -1 \end{bmatrix}$ $D = \begin{bmatrix} 9 & 0 & 0 \\ 0 & 6 & 0 \\ 0 & 0 & 2 \end{bmatrix}$

15. $P = \begin{bmatrix} 1 & 1 & 0 & 0 \\ 0 & 0 & 1 & -1 \\ 1 & -1 & 0 & 0 \\ 0 & 0 & 1 & 1 \end{bmatrix}$ $D = \begin{bmatrix} 2 & 0 & 0 & 0 \\ 0 & 2 & 0 & 0 \\ 0 & 0 & 2 & 0 \\ 0 & 0 & 0 & 0 \end{bmatrix}$

16. $P = \begin{bmatrix} 1 & 0 & -2 & 0 \\ 2 & 0 & 1 & 0 \\ 0 & 1 & 0 & -1 \\ 0 & 1 & 0 & 0 \end{bmatrix}$ $D = \begin{bmatrix} 5 & 0 & 0 & 0 \\ 0 & 4 & 0 & 0 \\ 0 & 0 & 0 & 0 \\ 0 & 0 & 0 & 0 \end{bmatrix}$

17. $P = \begin{bmatrix} 1 & 0 & 1 & 0 \\ 0 & 1 & 0 & -1 \\ 1 & 0 & -1 & 0 \\ 0 & 1 & 0 & 1 \end{bmatrix}$ $D = \begin{bmatrix} 2 & 0 & 0 & 0 \\ 0 & 2 & 0 & 0 \\ 0 & 0 & 0 & 0 \\ 0 & 0 & 0 & 0 \end{bmatrix}$

18. $P = \begin{bmatrix} 1 & 0 & 1 & 0 \\ 0 & 1 & 0 & -1 \\ 1 & 0 & -1 & 0 \\ 0 & 1 & 0 & 1 \end{bmatrix}$ $D = \begin{bmatrix} 4 & 0 & 0 & 0 \\ 0 & 4 & 0 & 0 \\ 0 & 0 & 2 & 0 \\ 0 & 0 & 0 & 0 \end{bmatrix}$

19. With $\lambda_1 = 3$ the matrix $A - \lambda I$ is

$$\begin{bmatrix} 1 & 1 & 0 & 0 & 0 \\ 1 & 1 & 0 & 0 & 0 \\ 0 & 0 & 1 & 0 & 1 \\ 0 & 0 & 0 & 2 & 0 \\ 0 & 0 & 1 & 0 & 0 \end{bmatrix}.$$

Hence two linearly independent eigenvectors associated with $\lambda_1 = 3$ are $\bar{v}_1 = (1,-1,0,0,0)$ and $\bar{v}_2 = (0,0,1,0,-1)$. With $\lambda_2 = 5$ the matrix $A - \lambda I$ is

$$\begin{bmatrix} -1 & 1 & 0 & 0 & 0 \\ 1 & -1 & 0 & 0 & 0 \\ 0 & 0 & -1 & 0 & 1 \\ 0 & 0 & 0 & 0 & 0 \\ 0 & 0 & 1 & 0 & -1 \end{bmatrix}.$$

Hence three linearly independent eigenvectors associated with $\lambda_2 = 5$ are $\bar{v}_3 = (1,1,0,0,0)$, $\bar{v}_4 = (0,0,1,0,1)$, and $\bar{v}_5 = (0,0,0,1,0)$.

20. The characteristic polynomial of the given matrix A is

$$(4 - \lambda)[(4 - \lambda)^2 - 4]\cdot(6 - \lambda)[(5 - \lambda)^2 - 1]$$

$$= (4 - \lambda)(2 - \lambda)(6 - \lambda)\cdot(6 - \lambda)(4 - \lambda)(6 - \lambda)$$

$$= (2 - \lambda)(4 - \lambda)^2(6 - \lambda)^3.$$

Hence A has a 1-dimensional eigenspace associated with $\lambda_1 = 2$, a 2-dimensional eigenspace associated with $\lambda_2 = 4$, and a 3-dimensional eigenspace associated with $\lambda_3 = 6$.

21. Because $|A| = |A^T|$, the fact that $AA^T = I$ implies that $|A|^2 = |AA^T| = 1$, so $|A| = \pm 1$.

22. Suppose A is orthogonal, so $A^{-1} = A^T$. Taking inverses, we get

$$(A^T)^{-1} = (A^{-1})^{-1} = A = (A^T)^T.$$

Thus the inverse of A^T equals its transpose, so A^T is orthogonal.

23. The matrix A is orthogonal if and only if $A^TA = I$. But the ijth element of A^TA is the dot product of the ith and jth column vectors of A. Hence $A^TA = I$ if and only if the columns of A are orthonormal.

24. The matrix A is orthogonal if and only if $AA^T = I$. But the ijth element of AA^T is the dot product of the ith and jth row vectors of A. Hence $AA^T = I$ if and only if

the rows of A are orthonormal.

25. If A and B are orthogonal, then $A^{-1} = A^T$ and $B^{-1} = B^T$, so

$$(AB)^{-1} = B^{-1}A^{-1} = B^TA^T = (AB)^T.$$

Thus AB is also orthogonal.

26. If $B = P^TAP$ and A is symmetric then

$$B^T = (P^TAP)^T = P^TA^TP = P^TAP = B$$

so B is symmetric.

27. If A is symmetric then $A = PDP^T$ where P is orthogonal and D is a diagonal matrix. Then $A^n = PD^nP^T = PEP^T$ where E is diagonal, so

$$(A^n)^T = (PEP^T)^T = PE^TP^T = PEP^T = A^n.$$

Thus A^n is symmetric.

28. If the $n \times n$ matrix A is diagonalizable then (by Theorem 1 in Section 6.2) it has n linearly independent eigenvectors. Hence the sum of the dimensions of the eigenspaces of A is n. Using the Gram-Schmidt algorithm we can construct an orthogonal basis for each of these eigenspaces. Because the eigenspaces of A are mutually orthogonal, the union of these orthogonal bases is a set of n mutually orthogonal eigenvectors of A. Therefore A is orthogonally diagonalizable by Theorem 3, and hence is symmetric by Theorem 4 in this section.

CHAPTER 7

LINEAR TRANSFORMATIONS

SECTION 7.1

MATRIX TRANSFORMATIONS

The purpose of this section is to introduce the concept of a linear transformation in a concrete way by means of examples of matrix transformations between Euclidean spaces. To this end we devote considerable discussion to the case of plane transformations whose geometric effects are readily visualized. Linear transformations between abstract vector spaces, and even such simple general facts as that $T(\bar{0}) = \bar{0}$ if T is linear, are deferred until Section 7.2.

Each transformation T defined in Problems 1, 2, 7, 8, 9, 10 is linear because there is a matrix A such that $T(\bar{x}) = A\bar{x}$. For each of these problems we simply give the matrix A.

1. $A = \begin{bmatrix} 1 & 1 \\ 1 & -1 \end{bmatrix}$

2. $A = \begin{bmatrix} 2 & 3 \\ 3 & -2 \end{bmatrix}$

3. Note that $T(1,0) = (1,3)$ and $T(2,0) = (2,3)$. Hence $T(2,0) \neq 2T(1,0)$, so T is not linear.

4. Note that $T(1,1) = (1,2)$ and $T(2,2) = (4,4)$. Hence $T(2,2) \neq 2T(1,1)$, so T is not linear.

5. Note that $T(0,0,1) = (0,0,1)$ and $T(0,0,2) = (0,0,8)$. Hence $T(0,0,2) \neq 2T(0,0,1)$, so T is not linear.

6. Note that $T(1,0,0) = (0,1,1)$ and $T(2,0,0) = (0,2,4)$. Hence $T(2,0,0) \neq 2T(1,0,0)$, so T is not linear.

7. $A = \begin{bmatrix} 1 & -3 \\ 3 & -1 \\ 2 & 5 \end{bmatrix}$

8. $A = \begin{bmatrix} 2 & 1 \\ 1 & 5 \\ 1 & -1 \end{bmatrix}$

9. $A = \begin{bmatrix} 2 & -3 & 4 \\ 3 & -5 & -7 \end{bmatrix}$

10. $A = \begin{bmatrix} 3 & 0 & -1 \\ 2 & -4 & 0 \end{bmatrix}$

11. $T(x,y) = (3x + 4y,\ 2x + 3y)$

12. $T(x,y) = (x - 3y,\ -2x + 2y)$

13. $T(x,y) = (x + 4y,\ 2x + 5y,\ 3x + 6y)$

14. $T(x,y,z) = (x + 4y + 5z,\ -x + 3y)$

15. $T(x,y,z) = (x + 3y,\ 2x + 5z,\ 4y + 6z)$

16. $T(u,v,x,y) = (2u + 3x,\ 4v + 5y)$

In each of Problems 17-22 we refer to Examples 6-8 in the text for the matrices of the prescribed geometric operations, and then multiply these matrices in the order corresponding to Equations (15) and (16).

17. $A = \begin{bmatrix} 1 & 0 \\ 0 & -1 \end{bmatrix}\begin{bmatrix} 2 & 0 \\ 0 & 1 \end{bmatrix} = \begin{bmatrix} 2 & 0 \\ 0 & -1 \end{bmatrix}$

18. $A = \begin{bmatrix} -1 & 0 \\ 0 & 1 \end{bmatrix}\begin{bmatrix} 1 & 0 \\ 0 & 3 \end{bmatrix} = \begin{bmatrix} -1 & 0 \\ 0 & 3 \end{bmatrix}$

19. $A = \begin{bmatrix} 0 & 1 \\ 1 & 0 \end{bmatrix}\begin{bmatrix} 0 & -1 \\ 1 & 0 \end{bmatrix} = \begin{bmatrix} 1 & 0 \\ 0 & -1 \end{bmatrix}$

20. $A = \begin{bmatrix} -1 & 0 \\ 0 & -1 \end{bmatrix}\begin{bmatrix} 0 & -1 \\ -1 & 0 \end{bmatrix} = \begin{bmatrix} 0 & 1 \\ 1 & 0 \end{bmatrix}$

21. $A = \begin{bmatrix} 0 & 1 \\ 1 & 0 \end{bmatrix}\begin{bmatrix} 1 & 0 \\ 0 & 3 \end{bmatrix}\begin{bmatrix} 2 & 0 \\ 0 & 1 \end{bmatrix} = \begin{bmatrix} 0 & 3 \\ 2 & 0 \end{bmatrix}$

22. $A = \begin{bmatrix} 1 & 0 \\ 0 & 2\sqrt{2} \end{bmatrix}\begin{bmatrix} \sqrt{2} & 0 \\ 0 & 1 \end{bmatrix}\begin{bmatrix} 1/\sqrt{2} & -1/\sqrt{2} \\ 1/\sqrt{2} & 1/\sqrt{2} \end{bmatrix} = \begin{bmatrix} 1 & -1 \\ 2 & 2 \end{bmatrix}$

23. $A = \begin{bmatrix} 1 & 0 \\ 0 & 3 \end{bmatrix}\begin{bmatrix} 2 & 0 \\ 0 & 1 \end{bmatrix}$

Thus T consists of an expansion by a factor of 2 in the x-direction, followed by an expansion by a factor of 3 in the y-direction.

24. $A = \begin{bmatrix} 2 & 0 \\ 0 & 1 \end{bmatrix}\begin{bmatrix} 1 & 0 \\ 0 & -1 \end{bmatrix}$

Thus T consists of a reflection in the x-axis, followed by an expansion by a factor of 2 in the x-direction.

25. $A = \begin{bmatrix} 0 & 1 \\ 1 & 0 \end{bmatrix}\begin{bmatrix} 3 & 0 \\ 0 & 1 \end{bmatrix}\begin{bmatrix} 1 & 0 \\ 0 & 2 \end{bmatrix}$

Thus T consists first of an expansion by a factor of 2 in the y-direction, then an expansion by a factor of 3 in the x-direction, and then a reflection in the line $y = x$.

26. $A = \begin{bmatrix} 0 & 1 \\ 1 & 0 \end{bmatrix}\begin{bmatrix} -1 & 0 \\ 0 & 1 \end{bmatrix}\begin{bmatrix} 1 & 0 \\ 0 & 3 \end{bmatrix}$

Thus T consists first of an expansion by a factor of 3 in the y-direction, then a reflection in the y-axis, and then a reflection in the line $y = x$.

27. $A = \begin{bmatrix} 1 & 0 \\ 1 & 1 \end{bmatrix}\begin{bmatrix} 1 & 1 \\ 0 & 1 \end{bmatrix}$

Hence T consists of a shear in the x-direction followed by a shear in the y-direction, each with $c = 1$.

28. $A = \begin{bmatrix} 1 & 0 \\ 2 & 1 \end{bmatrix}\begin{bmatrix} 1 & 3 \\ 0 & 1 \end{bmatrix}$

Hence T consists of a shear in the x-direction with $c = 3$, followed by a shear in the y-direction with $c = 2$.

29. $A = \begin{bmatrix} 1 & 0 \\ 2 & 1 \end{bmatrix}\begin{bmatrix} 1 & 0 \\ 0 & 3 \end{bmatrix}\begin{bmatrix} 1 & 2 \\ 0 & 1 \end{bmatrix}$

Hence T consists first of a shear in the x-direction with $c = 2$, then an expansion by a factor of 3 in the y-direction, and then a shear in the y-direction with $c = 2$.

30. $A = \begin{bmatrix} 1 & 0 \\ 2 & 1 \end{bmatrix}\begin{bmatrix} 0 & 1 \\ 1 & 0 \end{bmatrix}\begin{bmatrix} 1 & 0 \\ 3 & 1 \end{bmatrix}$

Hence T consists first of a shear in the y-direction with $c = 3$, then a reflection in the line $y = x$, and then a shear in the y-direction with $c = 2$.

31. $\begin{bmatrix} -1 & 0 \\ 0 & 1 \end{bmatrix}\begin{bmatrix} 1 & 0 \\ 0 & -1 \end{bmatrix} = \begin{bmatrix} -1 & 0 \\ 0 & -1 \end{bmatrix}$

32. $\begin{bmatrix} 0 & -1 \\ -1 & 0 \end{bmatrix}\begin{bmatrix} 0 & 1 \\ 1 & 0 \end{bmatrix} = \begin{bmatrix} -1 & 0 \\ 0 & -1 \end{bmatrix}$

33. $\begin{bmatrix} \cos\theta & -\sin\theta \\ \sin\theta & \cos\theta \end{bmatrix}\begin{bmatrix} 1 & 0 \\ 0 & -1 \end{bmatrix}\begin{bmatrix} \cos\theta & \sin\theta \\ -\sin\theta & \cos\theta \end{bmatrix}$

$$= \begin{bmatrix} \cos\theta & -\sin\theta \\ \sin\theta & \cos\theta \end{bmatrix}\begin{bmatrix} \cos\theta & \sin\theta \\ \sin\theta & -\cos\theta \end{bmatrix}$$

$$= \begin{bmatrix} \cos^2\theta - \sin^2\theta & 2\sin\theta\cos\theta \\ 2\sin\theta\cos\theta & \sin^2\theta - \cos^2\theta \end{bmatrix} = \begin{bmatrix} \cos 2\theta & \sin 2\theta \\ \sin 2\theta & -\cos 2\theta \end{bmatrix}$$

34. $$\begin{bmatrix} \cos\theta & \sin\theta \\ \sin\theta & -\cos\theta \end{bmatrix}\begin{bmatrix} 1 & 0 \\ 0 & -1 \end{bmatrix}\begin{bmatrix} \cos\theta & -\sin\theta \\ \sin\theta & \cos\theta \end{bmatrix}$$

By Problem 33, the first factor on the left represents a reflection in the line through the origin making an angle of $\theta/2$ with the positive x-axis, and the matrix on the right represents a counterclockwise rotation through an angle of θ.

35. $$\begin{bmatrix} \cos 2\beta & \sin 2\beta \\ \sin 2\beta & -\cos 2\beta \end{bmatrix}\begin{bmatrix} \cos 2\alpha & \sin 2\alpha \\ \sin 2\alpha & -\cos 2\alpha \end{bmatrix}$$

$$= \begin{bmatrix} \cos 2\beta\cos 2\alpha + \sin 2\beta\sin 2\alpha & \cos 2\beta\sin 2\alpha - \sin 2\beta\cos 2\alpha \\ \sin 2\beta\cos 2\alpha - \cos 2\beta\sin 2\alpha & \sin 2\beta\sin 2\alpha + \cos 2\beta\cos 2\alpha \end{bmatrix}$$

$$= \begin{bmatrix} \cos(2\beta - 2\alpha) & -\sin(2\beta - 2\alpha) \\ \sin(2\beta - 2\alpha) & \cos(2\beta - 2\alpha) \end{bmatrix}$$

By Problem 33, the original two factors represent reflections in the lines L_1 and L_2, whereas the final result represents a counterclockwise rotation through an angle of $2\beta - 2\alpha = 2(\beta - \alpha)$.

SECTION 7.2

PROPERTIES OF LINEAR TRANSFORMATIONS

This section concentrates on those aspects of the elementary theory of linear transformations that are needed to discuss whether a given linear transformation $T:V \longrightarrow W$ is an isomorphism -- whether it is both one-to-one and onto. The kernel and the range of T measure the extent to which a linear transformation fails to be one-to-one and onto. The key theoretical result in this section is the fact (Theorem 3) that, if V is finite-dimensional, then the dimension of V equals the sum of the dimensions of the kernel and the range of T.

1. $T(0,0) = (1,1) \neq (0,0)$

2. $T(0,0) = (0,-1) \neq (0,0)$

3. $T(0,0) = T(-1,-1) = (1,2)$, so $T(-1,-1) \neq -T(1,1)$

4. $T(1,1) = T(-1,-1) = (1,1)$, so $T(-1,-1) \neq -T(1,1)$

In each of Problems 5-10 we let $V = [\bar{v}_1 \ \bar{v}_2]$ and $W = [\bar{w}_1 \ \bar{w}_2]$. Then we want $AV = W$, so we take the matrix $A = WV^{-1}$. The 2×2 matrices involved are readily invertible, and we list for each problem only the desired matrix A.

5. $A = \begin{bmatrix} -1 & 1 \\ -3 & 2 \end{bmatrix}$

6. $A = \begin{bmatrix} -3 & 2 \\ -4 & 3 \end{bmatrix}$

7. $A = \begin{bmatrix} 0 & -1 \\ 2 & 3 \end{bmatrix}$

8. $A = \begin{bmatrix} 10 & -7 \\ -1 & 1 \end{bmatrix}$

9. $A = \begin{bmatrix} 26 & -11 \\ 8 & -3 \end{bmatrix}$

10. $A = \begin{bmatrix} -3 & 3 \\ 20 & -26 \end{bmatrix}$

11. $A = \begin{bmatrix} -5 & 9 \\ -8 & 13 \\ 18 & -29 \end{bmatrix}$

12. $A = \begin{bmatrix} -9 & 12 \\ -10 & 13 \\ -11 & 14 \end{bmatrix}$

In each of Problems 13-24 the method is to reduce A to an echelon matrix E. Then the rank r of T is the number of pivot columns of E, and the dimension of the kernel of T is $k = n - r$.

13. $E = \begin{bmatrix} 2 & 1 \\ 0 & 1 \end{bmatrix}$ $r = 2$ $k = 0$

14. $E = \begin{bmatrix} 3 & -2 \\ 0 & 1 \end{bmatrix}$ $r = 1$ $k = 1$

15. $E = \begin{bmatrix} 3 & 1 & -2 \\ 0 & 0 & 0 \end{bmatrix}$ $r = 1$ $k = 2$

16. $E = \begin{bmatrix} 1 & 0 & 6 \\ 0 & 1 & 1 \end{bmatrix}$ $r = 2$ $k = 1$

17. $E = \begin{bmatrix} 1 & 0 \\ 0 & 1 \\ 0 & 0 \end{bmatrix}$ $r = 2$ $k = 0$

18. $E = \begin{bmatrix} 1 & 2 \\ 0 & 0 \\ 0 & 0 \end{bmatrix}$ $r = 1$ $k = 1$

19. $E = \begin{bmatrix} 1 & 0 & 0 \\ 0 & 1 & 0 \\ 0 & 0 & 1 \end{bmatrix}$ $r = 3$ $k = 0$

20. $E = \begin{bmatrix} 1 & 0 & 5 \\ 0 & 1 & -1 \\ 0 & 0 & 0 \end{bmatrix}$ $r = 2$ $k = 1$

21. $E = \begin{bmatrix} 1 & 0 & 1 & 4 \\ 0 & 1 & 0 & 2 \\ 0 & 0 & 0 & 0 \end{bmatrix}$ $r = 2$ $k = 2$

22. $E = \begin{bmatrix} 1 & 0 & 1 & 0 \\ 0 & 1 & 2 & 0 \\ 0 & 0 & 0 & 1 \end{bmatrix}$ $r = 3$ $k = 1$

23. $E = \begin{bmatrix} 1 & 0 & 0 & 6 \\ 0 & 1 & 0 & -4 \\ 0 & 0 & 1 & 3 \end{bmatrix}$ $\qquad r = 3 \qquad k = 1$

24. $E = \begin{bmatrix} 3 & 0 & 0 & 16 & -9 \\ 0 & 4 & 0 & -9 & 16 \\ 0 & 0 & 12 & 11 & -60 \end{bmatrix}$ $\qquad r = 3 \qquad k = 2$

25. $T(u,v) = (x,y)$ is defined by $x = 2u + 4v$ and $y = 3u + 6v$. Obviously $3x = 2y$, so the kernel of T is the line $3x - 2y = 0$. The range of T is the line $2x + 4y = 3x + 6y = 0$, that is, $x + 2y = 0$.

26. Obviously the third row of A is the sum of the first two rows, so it follows that the range of T is the plane $z = x + y$. To find the kernel of T we solve the homogeneous system $A\bar{x} = \bar{0}$, and obtain $x = 4t$, $y = -3t$, $z = t$. These are the parametric equations of the straight line through the origin in R^3 that is the kernel of T.

27. Given vectors $\bar{w}_1$ and $\bar{w}_2$ in W, and scalars c_1 and c_2, we want to show that

$$T^{-1}(c_1\bar{w}_1 + c_2\bar{w}_2) = c_1T^{-1}(\bar{w}_1) + c_2T^{-1}(\bar{w}_2),$$

that is, that

$$T^{-1}(c_1\bar{w}_1 + c_2\bar{w}_2) = c_1\bar{v}_1 + c_2\bar{v}_2,$$

where $\bar{v}_1$ and $\bar{v}_2$ are the unique vectors in V such that $T(\bar{v}_i) = \bar{w}_i$ for $i = 1,2$. But this follows immediately from the fact that

$$T(c_1\bar{v}_1 + c_2\bar{v}_2) = c_1T(\bar{v}_1) + c_2T(\bar{v}_2) = c_1\bar{w}_1 + c_2\bar{w}_2$$

by linearity of T, and the fact that T is one-to-one.

28. Let V and W have bases $\{\bar{v}_1, \bar{v}_2, \cdots, \bar{v}_n\}$ and

$\{\bar{w}_1, \bar{w}_2, \cdots, \bar{w}_3\}$, respectively. Then, since vectors in V and W are uniquely expressible as linear combinations of these basis vectors, it follows readily that the function $T:V \longrightarrow W$ defined by

$$T(c_1\bar{v}_1 + \cdots + c_n\bar{v}_n) = c_1\bar{w}_1 + \cdots + c_n\bar{w}_n$$

is an isomorphism of V onto W.

In each of Problems 29-31 we denote by r the rank of T and by k the dimension of the kernel of T. Then $r + k = n$ by Theorem 3.

29. If $m < n$ then $r \leq m < n$ and $r + k = n$ imply that $k > 0$. Thus the kernel of T is nonzero, so it follows that T is not one-to-one.

30. If $n < m$ then $r \leq r + k = n < m$, so the dimension of the range of T is less than the dimension of W. Thus T cannot be unto W.

31. If $n = m$ then $r + k = m$, so $r = m$ if and only if $k = 0$. But T is one-to-one if $k = 0$, and T is onto if and only if $r = m$.

32. If A is not invertible and thus is a singular matrix, then the homogenous linear system $T(\bar{x}) = A\bar{x} = \bar{0}$ has a nontrivial solution $\bar{x}$, so T is not one-to-one and therefore is not an isomorphism. If the matrix A is invertible, then for each $\bar{y}$ the system $T(\bar{x}) = A\bar{x} = \bar{y}$ has the unique solution $\bar{x} = A^{-1}\bar{y}$, so T is both one-to-one and onto, and is therefore an isomorphism.

33. Suppose that the vectors $\bar{v}_1, \bar{v}_2, \cdots, \bar{v}_k$ in V are linearly independent, and that

$$c_1T(\bar{v}_1) + c_2T(\bar{v}_2) + \cdots + c_kT(\bar{v}_k) = \bar{0}.$$

Then by linearity it follows that

$$T(c_1\bar{v}_1 + c_2\bar{v}_2 + \cdots + c_k\bar{v}_k) = \bar{0},$$

which, because T is one-to-one, implies that

$$c_1\bar{v}_1 + c_2\bar{v}_2 + \cdots + c_k\bar{v}_k = \bar{0}.$$

Finally the fact that the vectors $\bar{v}_1, \bar{v}_2, \cdots, \bar{v}_k$ are linearly independent now implies that $c_1 = c_2 = \cdots = c_k = 0$, so we have shown that the image vectors are linearly independent.

34. In the notation of the proof of Theorem 3, let $\{\bar{v}_1, \bar{v}_2 \cdots, \bar{v}_n\}$ be a basis for V such that $\{\bar{v}_1, \bar{v}_2, \cdots, \bar{v}_k\}$ is a basis for $K = \text{Ker}(T)$, and then let U be the subspace of V that is generated by the vectors $\bar{v}_{k+1}, \cdots, \bar{v}_n$. Then it is immediate that $K + U = V$ and that K and U intersect in the zero vector. Moreover, we saw in the proof of Theorem 3 that the vectors $T(\bar{v}_{k+1}), \cdots, T(\bar{v}_n)$ span the range R of T, so it follows that T maps U onto R. Finally, to prove that T is one-to-one on U, let $\bar{u}_1$ and $\bar{u}_2$ be vectors in U such that

$$T(\bar{u}_1) = T(\bar{u}_2), \quad \text{so} \quad T(\bar{u}_1 - \bar{u}_2) = \bar{0}.$$

Then $\bar{u}_1 - \bar{u}_2$ lies both in U and in K, so it follows that $\bar{u}_1 = \bar{u}_2$. Thus no two different vectors in U can have the same image under T.

SECTION 7.3

COORDINATES AND CHANGE OF BASIS

In Problems 1-8 we are given a vector $\bar{v}$ in standard coordinates and want to change to coordinates with respect to the basis B. Equation (12) in the text says that

$$I\bar{v} = M\bar{v}_B$$

where $M = [\bar{v}_1 \ \bar{v}_2]$. Hence we need only compute

$$\bar{v}_B = M^{-1}\bar{v},$$

and therefore give only the answer for each of these problems.

1. $\bar{v}_B = (1,2)$ 2. $\bar{v}_B = (-2,1)$

3. $\bar{v}_B = (5,-5)$ 4. $\bar{v}_B = (4,-2)$

5. $\bar{v}_B = (1,-1)$ 6. $\bar{v}_B = (2,2)$

7. $\bar{v}_B = (-57,-4,18)$ 8. $\bar{v}_B = (-39,20,-7)$

In Problems 9-12 we use Equation (12) in the text, which tells us that

$$\begin{bmatrix} x \\ y \end{bmatrix} = [\bar{u}_1\ \bar{u}_2] \begin{bmatrix} x' \\ y' \end{bmatrix}.$$

Hence we need only carry out the matrix multiplication on the right.

9. $x = 3x' + y', \qquad y = x' + 4y'$

10. $x = x' - 3y', \qquad y = 2x' + 4y'$

11. $x = -x' - 3y', \qquad y = 2x' - 2y'$

12. $x = 2x' + 5y', \qquad y = -5x' + 2y'$

In Problems 13-16 it follows from Equation (12) that

$$\begin{bmatrix} x_1 \\ x_2 \end{bmatrix} = (M_B)^{-1}M_C \begin{bmatrix} y_1 \\ y_2 \end{bmatrix}$$

where $M_B = [\bar{u}_1\ \bar{u}_2]$ and $M_C = [\bar{v}_1\ \bar{v}_2]$.

13. $x_1 = y_1 + y_2, \qquad x_2 = y_1 + 2y_2$

14. $x_1 = 4y_1 - 3y_2, \qquad x_2 = -6y_1 + 6y_2$

15. $x_1 = 3y_1 - y_2, \qquad x_2 = y_1 + 3y_2$

16. $x_1 = 3y_1 + y_2, \qquad x_2 = -y_1 + 3y_2$

17. $x_1 = y_1 + 2y_2 + 4y_3, \qquad x_2 = 3y_1 - y_2 + 3y_3,$
$x_3 = 2y_1 + 3y_2 + 2y_3$

18. $x_1 = 38y_1 - 60y_2 - y_3, \qquad x_2 = 2y_1 - 4y_2 + y_3,$
$x_3 = -11y_1 + 19y_2 + y_3$

In each of Problems 19-26 we give first the eigenvalues and associated eigenvectors of the matrix A, and then the resulting geometric description of the transformation T.

19. $\lambda_1 = 5 : \bar{v}_1 = (1,1)$ and $\lambda_2 = 3 : \bar{v}_2 = (1,-1)$
T consists of an expansion of 5 in the direction of (1,1), and an expansion of 3 in the direction of (1,-1).

20. $\lambda_1 = 7 : \bar{v}_1 = (1,1)$ and $\lambda_2 = 3 : \bar{v}_2 = (1,-1)$
T consists of an expansion of 7 in the direction of (1,1), and an expansion of 3 in the direction of (1,-1).

21. $\lambda_1 = 10 : \bar{v}_1 = (2,-1)$ and $\lambda_2 = 5 : \bar{v}_2 = (1,2)$
T consists of an expansion of 10 in the direction of (2,-1), and an expansion of 5 in the direction of (1,2).

22. $\lambda_1 = 15 : \bar{v}_1 = (2,-1)$ and $\lambda_2 = 5 : \bar{v}_2 = (1,2)$
T consists of an expansion of 15 in the direction of (2,-1), and an expansion of 5 in the direction of (1,2).

23. $\lambda_1 = 9 : \bar{v}_1 = (1,1,1), \quad \lambda_2 = 6 : \bar{v}_2 = (1,-2,1),$
$\lambda_3 = 0 : \bar{v}_3 = (1,0,-1)$
T consists of an expansion of 9 in the direction of (1,1,1), an expansion of 6 in the direction of (1,-2,1), and then an orthogonal projection into the plane $x - z = 0$ through the origin that is orthogonal to the vector (1,0,-1).

24. $\lambda_1 = 9 : \bar{v}_1 = (1,2,2), \quad \lambda_2 = 6 : \bar{v}_2 = (0,1,-1),$
$\lambda_3 = 0 : \bar{v}_3 = (4,-1,-1)$

T consists of an expansion of 9 in the direction of (1,2,2), an expansion of 6 in the direction of (0,1,-1), and then an orthogonal projection into the plane $4x - y - z = 0$ through the origin that is orthogonal to the vector (4,-1,-1).

25. $\lambda_1 = 6 : \bar{v}_1 = (1,1,1)$, $\lambda_2 = 3 : \bar{v}_2 = (1,-2,1)$ and $\bar{v}_3 = (1,0,-1)$

T consists of an expansion of 6 in the direction of (1,1,1), plus expansions of 3 in the directions of both (1,-2,1) and (1,0,-1).

26. $\lambda_1 = 9 : \bar{v}_1 = (1,1,1)$, $\lambda_2 = 6 : \bar{v}_2 = (1,-2,1)$, $\lambda_3 = 2 : \bar{v}_3 = (1,0,-1)$

T consists of an expansion of 9 in the direction of (1,1,1), an expansion of 6 in the direction of (1,-2,1), and an expansion of 2 in the direction of (1,0,-1).

By Theorem 2, the matrix A in Problems 27 and 28 is given by

$$A = [T(\bar{v}_1)_B \quad T(\bar{v}_2)_B] = M^{-1}[T(\bar{v}_1) \quad T(\bar{v}_2)]$$

where $M = [\bar{v}_1 \quad \bar{v}_2]$.

27. $$A = \begin{bmatrix} 2 & -1 \\ -1 & 1 \end{bmatrix}^{-1} \begin{bmatrix} 1 & 0 \\ 3 & -2 \end{bmatrix} = \begin{bmatrix} 4 & -2 \\ 7 & -4 \end{bmatrix}$$

28. $$A = \begin{bmatrix} 2 & 1 \\ 3 & 2 \end{bmatrix}^{-1} \begin{bmatrix} 7 & 4 \\ 8 & 5 \end{bmatrix} = \begin{bmatrix} 6 & 3 \\ -5 & -2 \end{bmatrix}$$

29. For any basis $B = \{\bar{v}_1, \bar{v}_2, \cdots, \bar{v}_n\}$,

$$T(a_1\bar{v}_1 + a_2\bar{v}_2 + \cdots + a_n\bar{v}_n)$$

$$= ka_1\bar{v}_1 + ka_2\bar{v}_2 + \cdots + ka_n\bar{v}_n.$$

Therefore the matrix of T relative to B is the $n \times n$ diagonal matrix with k in each diagonal position.

30. (a) $D(1) = 0$, $D(e^x)$, and $D(xe^x) = e^x + xe^x$, so

$$A = \begin{bmatrix} 0 & 0 & 0 \\ 0 & 1 & 1 \\ 0 & 0 & 1 \end{bmatrix}.$$

(b) $D(e^x) = e^x$, $D(xe^x) = e^x + xe^x$, and

$D(x^2e^x) = 2xe^x + x^2e^x$, so

$$A = \begin{bmatrix} 1 & 1 & 0 \\ 0 & 1 & 2 \\ 0 & 0 & 1 \end{bmatrix}.$$

31. Since $J(1) = x$, $J(x) = (1/2)x^2$, and $J(x^2) = (1/3)x^3$, we see that

$$A_J = \begin{bmatrix} 0 & 0 & 0 \\ 1 & 0 & 0 \\ 0 & 1/2 & 0 \\ 0 & 0 & 1/3 \end{bmatrix}.$$

The fact that

$$A_DA_J = \begin{bmatrix} 0 & 1 & 0 & 0 \\ 0 & 0 & 2 & 0 \\ 0 & 0 & 0 & 3 \end{bmatrix} \begin{bmatrix} 0 & 0 & 0 \\ 1 & 0 & 0 \\ 0 & 1/2 & 0 \\ 0 & 0 & 1/3 \end{bmatrix} = \begin{bmatrix} 1 & 0 & 0 \\ 0 & 1 & 0 \\ 0 & 0 & 1 \end{bmatrix}$$

is consistent with the fact that the derivative of the integral of a polynomial is the same polynomial.

32. Let $\bar{x}$ and $\bar{x}'$ be the coordinate vectors of $\bar{v}$ with respect to the bases B and B'. Since $\bar{x} = P\bar{x}'$ and $\bar{x}' = Q\bar{x}$, it follows that

$$\bar{x} = P(Q\bar{x}) = PQ\bar{x}.$$

The fact that this is true for all $\bar{x}$ implies that $PQ = I$, so $P^{-1} = Q$.

SECTION 7.4

ISOMETRIES, ROTATIONS AND COMPUTER GRAPHICS

1. Note first that

$$|\bar{u} + \bar{v}|^2 = (\bar{u} + \bar{v})\cdot(\bar{u} + \bar{v}) = \bar{u}\cdot\bar{u} + 2\bar{u}\cdot\bar{v} + \bar{v}\cdot\bar{v},$$

so

$$\bar{u}\cdot\bar{v} = (1/2)[|\bar{u} + \bar{v}|^2 - |\bar{u}|^2 - |\bar{v}|^2].$$

If T preserves distances then it follows that

$$T\bar{u}\cdot T\bar{u} = (1/2)[|T\bar{u} + T\bar{v}|^2 - |T\bar{u}|^2 - |T\bar{v}|^2]$$

$$= (1/2)[|\bar{u} + \bar{v}|^2 - |\bar{u}|^2 - |\bar{v}|^2 = \bar{u}\cdot\bar{v},$$

so T also preserves dot products.

2. $\bar{u}\cdot A\bar{v} = \bar{u}^T A\bar{v} = (A^T\bar{u})^T\bar{v} = (A^T\bar{u})\cdot\bar{v}$

3. (a) If $T(\bar{x}) = A\bar{x}$ is an isometry, then

$$\begin{aligned} \bar{x}\cdot\bar{y} &= A\bar{x}\cdot A\bar{y} && \text{(by Problem 1)} \\ &= [A^T(A\bar{x})]\cdot\bar{y} && \text{(by Problem 2)} \\ &= (A^TA)\bar{x}\cdot\bar{y} \end{aligned}$$

(b) $A^TA\bar{e}_i\cdot\bar{e}_j = \bar{e}_i\cdot\bar{e}_j$ is the ijth element of the matrix A^TA. Since $\bar{e}_i\cdot\bar{e}_j$ is 1 if $i = j$ and 0 if $i \neq j$, it follows that $A^TA = I$. Thus A is orthogonal.

4. If T denotes the reflection in the y-axis then

$$F(\bar{x}) = \bar{e}_1 + T(\bar{x} - \bar{e}_1),$$

so

$$F(\bar{x}) = \begin{bmatrix} 1 \\ 0 \end{bmatrix} + \begin{bmatrix} -1 & 0 \\ 0 & 1 \end{bmatrix}\begin{bmatrix} x - 1 \\ y \end{bmatrix}$$

$$= \begin{bmatrix} 2 - x \\ y \end{bmatrix} = \begin{bmatrix} -1 & 0 \\ 0 & 1 \end{bmatrix}\begin{bmatrix} x \\ y \end{bmatrix} + \begin{bmatrix} 2 \\ 0 \end{bmatrix}.$$

Thus

$$A = \begin{bmatrix} -1 & 0 \\ 0 & 1 \end{bmatrix} \quad \text{and} \quad \bar{b} = \begin{bmatrix} 2 \\ 0 \end{bmatrix}.$$

5. $$F(\bar{x}) = \begin{bmatrix} 1 \\ -1 \end{bmatrix} + \begin{bmatrix} 0 & 1 \\ 1 & 0 \end{bmatrix}\begin{bmatrix} x - 1 \\ y + 1 \end{bmatrix}$$

$$= \begin{bmatrix} 0 & 1 \\ 1 & 0 \end{bmatrix}\begin{bmatrix} x \\ y \end{bmatrix} + \begin{bmatrix} 2 \\ -2 \end{bmatrix},$$

so

$$A = \begin{bmatrix} 0 & 1 \\ 1 & 0 \end{bmatrix} \quad \text{and} \quad \bar{b} = \begin{bmatrix} 2 \\ -2 \end{bmatrix}.$$

6. $$F(\bar{x}) = \begin{bmatrix} 2 \\ 3 \end{bmatrix} + \begin{bmatrix} 0 & -1 \\ 1 & 0 \end{bmatrix}\begin{bmatrix} x - 2 \\ y - 3 \end{bmatrix}$$

$$= \begin{bmatrix} 0 & -1 \\ 1 & 0 \end{bmatrix} + \begin{bmatrix} 5 \\ 1 \end{bmatrix},$$

so

$$A = \begin{bmatrix} 0 & -1 \\ 1 & 0 \end{bmatrix} \quad \text{and} \quad \bar{b} = \begin{bmatrix} 5 \\ 1 \end{bmatrix}.$$

7. $$F(\bar{x}) = \begin{bmatrix} 5 \\ 5 \end{bmatrix} + \begin{bmatrix} 4/5 & -3/5 \\ 3/5 & 4/5 \end{bmatrix}\begin{bmatrix} x - 5 \\ y - 5 \end{bmatrix}$$

$$= \begin{bmatrix} 4/5 & -3/5 \\ 3/5 & 4/5 \end{bmatrix}\begin{bmatrix} x \\ y \end{bmatrix} + \begin{bmatrix} 4/5 & -3/5 \\ 3/5 & 4/5 \end{bmatrix}\begin{bmatrix} -5 \\ -5 \end{bmatrix} + \begin{bmatrix} 5 \\ 5 \end{bmatrix},$$

so

$$A = \begin{bmatrix} 4/5 & -3/5 \\ 3/5 & 4/5 \end{bmatrix} \quad \text{and} \quad \bar{b} = \begin{bmatrix} 4 \\ -2 \end{bmatrix}.$$

8. If the isometry T of R^2 is a rotation, then Problem 34 in Section 7.1 says that T is the composition of two reflections. Otherwise, Theorem 2 in this section implies that $T = R_\theta R_x$, where R_x denotes the reflection in the x-axis, and R_θ is a rotation. But then Problem 34 in Section 7.1 says that $R_\theta = R_L R_x$, where R_L is a reflection in a line P through the origin. Hence

$$T = R_\theta R_x = R_L R_x R_x = R_L,$$

because $R_x R_x$ is the identity transformation. Thus in this case T is the single reflection R_L.

9. The matrix of the inversion T is

$$\begin{bmatrix} -1 & 0 & 0 \\ 0 & -1 & 0 \\ 0 & 0 & -1 \end{bmatrix} = \begin{bmatrix} 1 & 0 & 0 \\ 0 & 1 & 0 \\ 0 & 0 & -1 \end{bmatrix} \begin{bmatrix} -1 & 0 & 0 \\ 0 & -1 & 0 \\ 0 & 0 & 1 \end{bmatrix}.$$

Hence T is equivalent to a rotation of 180° about the z-axis, followed by a reflection in the xy-plane.

10. The transformation F can be carried out by translating the point $\bar{a}$ to the origin, rotating about the line through the origin, and then translating the origin back to $\bar{a}$. Hence

$$F(\bar{x}) = \bar{a} + A(\bar{x} - \bar{a}) = A\bar{x} + \bar{a} - A\bar{a},$$

so

$$\bar{b} = \bar{a} - A\bar{a} = (I - A)\bar{a}.$$

11. $A = R_z(45^\circ)R_x(45^\circ)R_z(-45^\circ)$

$$= \begin{bmatrix} (2+\sqrt{2})/4 & (2-\sqrt{2})/4 & 1/2 \\ (2-\sqrt{2})/4 & (2+\sqrt{2})/4 & -1/2 \\ -1/2 & 1/2 & 1/\sqrt{2} \end{bmatrix}$$

12. A rotation of -45° about the z-axis moves the line L to the line $z = x$ in the xz-plane, and then a rotation of -45° about the y-axis moves the latter line to the x-axis. Hence the matrix of T is

$$A = R_z(45^{\circ})R_y(45^{\circ})R_x(45^{\circ})R_y(-45^{\circ})R_z(-45^{\circ})$$

$$\approx \begin{bmatrix} 0.7803 & -0.4268 & 0.4571 \\ 0.5732 & 0.7803 & -0.2500 \\ -0.2500 & 0.4571 & 0.8536 \end{bmatrix}$$

13. Because $A' = P^{-1}AP$ with A and P orthogonal (so $A^T = A^{-1}$ and $P^T = P^{-1}$) it follows that

$$(A')^T = (P^TAP)^T = P^TA^TP = P^{-1}A^{-1}P = (A')^{-1},$$

so A' is orthogonal also.

14. The fact that A' is orthogonal (Problem 13) implies that

$$(A')^T = \begin{bmatrix} \pm 1 & 0 \\ 0 & (B')^T \end{bmatrix} = \begin{bmatrix} \pm 1 & 0 \\ 0 & (B')^{-1} \end{bmatrix} = (A')^{-1}.$$

Hence $(B')^T = (B')^{-1}$, so B' is orthogonal.

15. We need only show that the matrix

$$B' = \begin{bmatrix} \cos\theta & \sin\theta \\ \sin\theta & -\cos\theta \end{bmatrix}$$

has eigenvalues +1 and -1. But this follows because the characteristic polynomial of B' is

$$|B' - \lambda I| = (\lambda - \cos\theta)(\lambda + \cos\theta) - (\sin\theta)^2$$

$$= \lambda^2 - [\cos^2\theta + \sin^2\theta]$$

$$= \lambda^2 - 1$$

$$= (\lambda - 1)(\lambda + 1).$$

16. $$\begin{bmatrix} 1 & 0 & 0 \\ 0 & 4/5 & -3/5 \\ 0 & 3/5 & 4/5 \end{bmatrix}\begin{bmatrix} 4/5 & 0 & -3/5 \\ 0 & 1 & 0 \\ 3/5 & 0 & 4/5 \end{bmatrix}\begin{bmatrix} 5 & 0 & 0 \\ 0 & 5 & 0 \\ 0 & 0 & 5 \end{bmatrix}$$

$$= \begin{bmatrix} 4 & 0 & -3 \\ -1.8 & 4 & -2.4 \\ 2.4 & 3 & 3.2 \end{bmatrix}$$

so the images in the xy-plane of the vertices P_1, P_2, and P_3 are $Q_1(4,-1.8)$, $Q_2(0,4)$, and $Q_3(-3,-2.4)$.

17. $$\begin{bmatrix} 1 & 0 & 0 \\ 0 & 4/5 & -3/5 \\ 0 & 3/5 & 4/5 \end{bmatrix}\begin{bmatrix} 4/5 & 0 & -3/5 \\ 0 & 1 & 0 \\ 3/5 & 0 & 4/5 \end{bmatrix}\begin{bmatrix} 0 & 25 & 25 & 0 \\ 0 & 0 & 50 & 50 \\ 0 & 0 & 0 & 0 \end{bmatrix}$$

$$= \begin{bmatrix} 0 & 20 & 20 & 0 \\ 0 & -9 & 31 & 40 \\ 0 & 12 & 42 & 30 \end{bmatrix}$$

so the images of the vertices P_1, P_2, P_3 and P_4 in the xy-plane are $Q_1(0,0)$, $Q_2(20,-9)$, $Q_3(20,31)$, and $Q_4(0,40)$.

18. $$R_x(30^\circ)R_y(45^\circ)\begin{bmatrix} -1 & 0 & 1 & 0 \\ 0 & 0 & 0 & 2 \\ 0 & 2 & 0 & 0 \end{bmatrix}$$

$$\approx \begin{bmatrix} -(\sqrt{2})/2 & -\sqrt{2} & (\sqrt{2})/2 & 0 \\ (\sqrt{2})/4 & -(\sqrt{2})/2 & -(\sqrt{2})/4 & \sqrt{3} \\ -(\sqrt{6})/4 & (\sqrt{6})/2 & (\sqrt{6})/4 & 1 \end{bmatrix}$$

so the images of the vertices P_1, P_2, P_3 and P_4 in the xy-plane are $Q_1(-0.7071, 0.3536)$, $Q_2(-1.4142, -0.7071)$, $Q_3(0.7071, -0.3536)$, and $Q_4(0, 1.7321)$.

19. The matrix of T is

$$A = \begin{bmatrix} 1 & 0 & 0 \\ 0 & 0 & -1 \\ 0 & 1 & 0 \end{bmatrix}\begin{bmatrix} 0 & 0 & -1 \\ 0 & 1 & 0 \\ 1 & 0 & 0 \end{bmatrix} = \begin{bmatrix} 0 & 0 & -1 \\ -1 & 0 & 0 \\ 0 & 1 & 0 \end{bmatrix}.$$

By reducing A to the echelon form

$$E = \begin{bmatrix} 1 & 0 & 1 \\ 0 & 1 & -1 \\ 0 & 0 & 0 \end{bmatrix}$$

we find the eigenvector $\bar{v} = (1,-1,-1)$ associated with eigenvalue $\lambda = 1$. The vector $\bar{u} = (1,1,0)$ is orthogonal to $\bar{v}$, and $\bar{w} = T(\bar{u}) = A\bar{u} = (0,-1,1)$. Now $\bar{u} \times \bar{w} = \bar{v}$, so $(\bar{u}, \bar{w}, \bar{v})$ is a right-handed triple. Finally, the angle θ between $\bar{u}$ and $T(\bar{u})$ is determined by

$$\cos\theta = \frac{\bar{u}\cdot\bar{w}}{|\bar{u}|\,|\bar{w}|} = -\frac{1}{2},$$

so $\theta = 120^{\circ}$. Therefore T is a right-handed rotation through the angle 120° about the axis determined by the vector $\bar{v} = (1,-1,-1)$.

20. The matrix of T is

$$A = R_x(45^{\circ})R_y(150^{\circ}) \approx \begin{bmatrix} -0.8660 & 0.0000 & -0.5000 \\ -0.3536 & 0.7071 & 0.6124 \\ 0.3536 & 0.7071 & -0.6124 \end{bmatrix}$$

By reducing A to the echelon form

$$E \approx \begin{bmatrix} 1 & 0 & 0.2679 \\ 0 & 1 & -2.4142 \\ 0 & 0 & 0 \end{bmatrix}$$

We find the eigenvector $\bar{v} = (0.2679, -2.4142, -1)$ associated with the eigenvalue $\lambda = 1$. Next observe that the vector $\bar{u} = (2.4142, 0.2679, 0)$ is orthogonal to $\bar{v}$, and

$$\bar{w} = T(\bar{u}) = A\bar{u} = (-2.0908, -0.6641, 1.0430).$$

Now $(\bar{u}, \bar{w}, \bar{v})$ is a righthanded triple, and the angle between $\bar{u}$ and $\bar{w}$ is $\theta \approx 152.33^\circ$. Therefore T is a right-handed rotation through the angle θ about the axis through the origin determined by the vector $\bar{v}$.

CHAPTER 8

FURTHER APPLICATIONS

SECTION 8.1

CONIC SECTIONS AND QUADRATIC FORMS

In each of Problems 1-6 we list the equation obtained by completing squares, the type of conic section, the transformed equation in x'y'-coordinates, and the origin of the new system.

1. $2(x - 2)^2 + (y - 3)^2 = 4$

   Ellipse: $2(x')^2 + (y')^2 = 4$; Origin (2,3)

2. $(x + 3)^2 + 3(y + 2)^2 = 3$

   Ellipse: $(x')^2 + 3(y')^2 = 3$; Origin (-3,-2)

3. $9(x - 1)^2 - 16(y + 1)^2 = 144$

   Hyperbola: $9(x')^2 - 16(y')^2 = 144$; Origin (1,-1)

4. $(y - 2)^2 = 4(x + 3)$

   Parabola: $(y')^2 = 4x'$; Origin (2,-3)

5. $2(x - 2)^2 + 3(y - 3)^2 = 0$

   $2(x')^2 + 3(y')^2 = 0$; the single point (2,3)

6. $4(x - 1)^2 - (y + 2)^2 = 0$

   Intersecting lines: $y' = \pm 2x'$; Origin (1,-2)

In each of Problems 7-20 we list the diagonalizing orthogonal matrix P, the transformed equation in the rotated $x'y'$-coordinate system, the type of conic section, and the angle θ of counterclockwise rotation from the xy-system to the $x'y'$-system.

7. $P = \frac{1}{\sqrt{2}}\begin{bmatrix} 1 & -1 \\ 1 & 1 \end{bmatrix}$ $4(x')^2 + 2(y')^2 = 1$

Ellipse, $\theta = \tan^{-1}(1) = 45^\circ$

8. $P = \frac{1}{\sqrt{2}}\begin{bmatrix} 1 & -1 \\ 1 & 1 \end{bmatrix}$ $4(x')^2 - 2(y')^2 = 1$

Hyperbola, $\theta = \tan^{-1}(1) = 45^\circ$

9. $P = \frac{1}{\sqrt{5}}\begin{bmatrix} 2 & -1 \\ 1 & 2 \end{bmatrix}$ $5(x')^2 = 20$; Two lines $x' = \pm 2$

$\theta = \tan^{-1}(1/2) = 26.57^\circ$

10. $P = \frac{1}{\sqrt{5}}\begin{bmatrix} 2 & -1 \\ 1 & 2 \end{bmatrix}$ $10(x')^2 + 5(y')^2 = 40$

Ellipse, $\theta = \tan^{-1}(1/2) = 26.57^\circ$

11. $P = \frac{1}{\sqrt{(10)}}\begin{bmatrix} 3 & -1 \\ 1 & 3 \end{bmatrix}$ $5(x')^2 - 5(y')^2 = 5$

Hyperbola, $\theta = \tan^{-1}(1/3) \approx 18.43^\circ$

12. $P = \frac{1}{\sqrt{(10)}}\begin{bmatrix} 3 & -1 \\ 1 & 3 \end{bmatrix}$ $20(x')^2 + 10(y')^2 = 40$

Ellipse, $\theta = \tan^{-1}(1/3) \approx 18.43^\circ$

13. $P = \frac{1}{\sqrt{(13)}}\begin{bmatrix} 3 & -2 \\ 2 & 3 \end{bmatrix}$ $26(x')^2 + 13(y')^2 = 26$

Ellipse, $\theta = \tan^{-1}(2/3) \approx 33.69^\circ$

14. $P = \frac{1}{\sqrt{(13)}}\begin{bmatrix} 3 & -2 \\ 2 & 3 \end{bmatrix}$ $13(x')^2 = 13$: Two lines $x' = \pm 1$

$\theta = \tan^{-1}(2/3) \approx 33.69^\circ$

15. $P = \frac{1}{5}\begin{bmatrix} 3 & -4 \\ 4 & 3 \end{bmatrix}$ $100(x')^2 + 25(y')^2 = 100$

Ellipse, $\theta = \tan^{-1}(4/3) \approx 53.13^\circ$

16. $P = \frac{1}{5}\begin{bmatrix} 3 & -4 \\ 4 & 3 \end{bmatrix}$ $25(x')^2 = 0$, Line $x' = 0$

$\theta = \tan^{-1}(4/3) \approx 53.13^\circ$

17. $P = \frac{1}{\sqrt{(17)}}\begin{bmatrix} 4 & -1 \\ 1 & 4 \end{bmatrix}$ $34(x')^2 + 17(y')^2 = 68$

Ellipse, $\theta = \tan^{-1}(1/4) \approx 14.04^\circ$

18. $P = \frac{1}{\sqrt{(13)}}\begin{bmatrix} 3 & -2 \\ 2 & 3 \end{bmatrix}$ $52(x')^2 + 13(y')^2 = 52$

Ellipse, $\theta = \tan^{-1}(2/3) \approx 33.69^\circ$

19. $P = \frac{1}{13}\begin{bmatrix} 12 & -5 \\ 5 & 12 \end{bmatrix}$ $169(x')^2 - 169(y')^2 = 0$,

Two lines: $y' = \pm x'$,

$\theta = \tan^{-1}(5/12) \approx 22.62^\circ$

20. $P = \frac{1}{13}\begin{bmatrix} 12 & -5 \\ 5 & 12 \end{bmatrix}$ $338(x')^2 + 169(y')^2 = 0$,

Single point $(0,0)$

In each of Problems 21-26 we give the orthogonal diagonalizing matrix P and the angle θ of rotation, the transformed equation in rotated $x'y'$-coordinates, and the final equation after an appropriate translation to identify the type of conic section.

21. $P = \frac{1}{5}\begin{bmatrix} 3 & -4 \\ 4 & 3 \end{bmatrix}$ $\theta = \tan^{-1}(4/3) \approx 53.13^\circ$

$25(x')^2 + 100(y')^2 - 50x' - 75 = 0$

$25(x' - 1)^2 + 100(y')^2 = 100$

$(x'')^2 + 4(y'')^2 = 4$ Ellipse

22. $P = \frac{1}{5}\begin{bmatrix} 3 & -4 \\ 4 & 3 \end{bmatrix}$ $\qquad \theta = \tan^{-1}(4/3) \approx 53.13^\circ$

$25(x')^2 + 50(y')^2 - 100x' - 100y' + 125 = 0$

$25(x' - 2)^2 + 50(y' - 1)^2 = 25$

$(x'')^2 + 2(y'')^2 = 1$ Ellipse

23. $P = \frac{1}{5}\begin{bmatrix} 3 & -4 \\ 4 & 3 \end{bmatrix}$ $\qquad \theta = \tan^{-1}(4/3) \approx 53.13^\circ$

$-25(x')^2 + 50(y')^2 + 100x' - 100y' -75 = 0$

$50(y' - 1)^2 - 25(x' - 2)^2 = 25$

$2(y'')^2 - (x'')^2 = 1$ Hyperbola

24. $P = \frac{1}{5}\begin{bmatrix} 3 & -4 \\ 4 & 3 \end{bmatrix}$ $\qquad \theta = \tan^{-1}(4/3) \approx 53.13^\circ$

$25(x')^2 - 150x' + 100y' + 245 = 0$

$25(x' - 3)^2 = -100(y' + 1/5)$

$(x'')^2 = -4y''$ Parabola

25. $P = \frac{1}{17}\begin{bmatrix} 15 & -8 \\ 8 & 15 \end{bmatrix}$ $\qquad \theta = \tan^{-1}(8/15) \approx 28.07^\circ$

$289(x')^2 - 289(y')^2 - 578x' = 0$

$(x' - 1)^2 - (y')^2 = 1$

$(x'')^2 - (y'')^2 = 1$ Hyperbola

26. $P = \frac{1}{13}\begin{bmatrix} 5 & -12 \\ 12 & 3 \end{bmatrix}$ $\qquad \theta = \tan^{-1}(12/5) \approx 67.38^\circ$

$$169(y')^2 - 169x' - 169 = 0$$

$$(y')^2 = x' + 1$$

$$(y'')^2 = x'' \qquad \text{Parabola}$$

27. In this case we can complete squares in x and y to get the equation

$$\alpha x^2 + \beta y^2 = F,$$

where α and β are both positive. Then the graph is an ellipse if $F > 0$, the single point $(0,0)$ if $F = 0$, the empty set if $F < 0$.

28. In this case we can complete squares in x and y to get the equation

$$\alpha x^2 - \beta y^2 = F,$$

where α and β are both positive. Then the graph is a hyperbola if $F \neq 0$, two intersecting lines if $F = 0$.

29. All four cases follow readily from the fact that completing the square in x yields the equation

$$ax^2 + ey + f = 0.$$

30. The rotation $x = Px'$ of this section yields the equation

$$\lambda_1(x')^2 + \lambda_2(y')^2 + d'x' + e'y' + f = 0,$$

where λ_1 and λ_2 are the eigenvalues of the matrix A of the associated quadratic form. Now

$$\lambda_1\lambda_2 = \det(P^{-1}AP) = \det(A) = ac - b^2.$$

Hence if $ac - b^2 > 0$ we have the ellipse of Problem 27;
if $ac - b^2 < 0$ we have the hyperbola of Problem 28;
if $ac - b^2 = 0$ we have the parabola of Problem 29.

31. We find that the matrix

$$A = \begin{bmatrix} 2929 & -1728 \\ -1728 & 1921 \end{bmatrix}$$

has characteristic equation

$$\lambda_2 - 4850\lambda + 2640625 = 0.$$

The eigenvalues $\lambda_1 = 625$ and $\lambda_2 = 4225$ have associated eigenvectors $\bar{v}_1 = (3,4)$ and $\bar{v}_2 = (-4,3)$. The orthogonal diagonalizing matrix

$$P = \frac{1}{5}\begin{bmatrix} 3 & -4 \\ 4 & 3 \end{bmatrix}$$

yields the transformed equation

$$625(x')^2 + 4225(y')^2 - 15000x' - 15625 = 0$$

in the rotated coordinate system. Simplification finally yields

$$25(x')^2 + 169(y')^2 - 600x' - 625 = 0,$$

$$25(x' - 12)^2 + 169(y')^2 = 4225,$$

$$\frac{(x' - 12)^2}{169} + \frac{(y')^2}{25} = 1,$$

the equation of the indicated ellipse.

SECTION 8.2

QUADRATIC FORMS AND EXTREME VALUES

1. $\lambda_1 = 2$, $\lambda_2 = -2$: Indefinite

2. $\lambda_1 = 4$, $\lambda_2 = 2$: Positive definite

3. $\lambda_1 = 6$, $\lambda_2 = 4$: Positive definite

4. $\lambda_1 = 5$, $\lambda_2 = -5$: Indefinite

5. $\lambda_1 = 10$, $\lambda_2 = 5$: Positive definite

6. $\lambda_1 = 10$, $\lambda_2 = -5$: Indefinite

7. $\lambda_1 = 4$, $\lambda_2 = 4$, $\lambda_3 = 2$: Positive definite

8. $\lambda_1 = 4$, $\lambda_2 = 2$, $\lambda_3 = -2$: Indefinite

9. $\lambda_1 = 6$, $\lambda_2 = 3$, $\lambda_3 = -3$: Indefinite

10. $\lambda_1 = 9$, $\lambda_2 = 6$, $\lambda_3 = 2$: Positive definite

11. $\Delta_1 = 0$, $\Delta_2 = -4$: Indefinite

12. $\Delta_1 = 3$, $\Delta_2 = 8$: Positive definite

13. $\Delta_1 = 5$, $\Delta_2 = 24$: Positive definite

14. $\Delta_1 = 3$, $\Delta_2 = -25$: Indefinite

15. $\Delta_1 = 9$, $\Delta_2 = 50$: Positive definite

16. $\Delta_1 = 7$, $\Delta_2 = -50$: Indefinite

17. $\Delta_1 = 3$, $\Delta_2 = 12$, $\Delta_3 = 32$: Positive definite

18. $\Delta_1 = 1$, $\Delta_2 = -8$, $\Delta_3 = -16$: Indefinite

19. $\Delta_1 = 1, \quad \Delta_2 = 3, \quad \Delta_3 = -54$: Indefinite

20. $\Delta_1 = 5, \quad \Delta_2 = 34, \quad \Delta_3 = 108$: Positive definite

21. $\Delta_1 = 2, \quad \Delta_2 = 8, \quad \Delta_3 = 14, \quad \Delta_4 = 21$:
Positive definite

22. $\Delta_1 = 2, \quad \Delta_2 = 6, \quad \Delta_3 = 9, \quad \Delta_4 = 42$:
Positive definite

23. $\Delta_1 = 1, \quad \Delta_2 = -3, \quad \Delta_3 = 5, \quad \Delta_4 = -7$:
Indefinite

24. $\Delta_1 = 4, \quad \Delta_2 = 15, \quad \Delta_3 = 54, \quad \Delta_4 = -378$:
Indefinite

In each of Problems 25-30 we list the diagonalizing orthogonal matrix P, the transformed equation in x'y'z-coordinates, the type of surface, and the angle θ of rotation.

25. $P = \frac{1}{\sqrt{2}}\begin{bmatrix} 1 & -1 \\ 1 & 1 \end{bmatrix}$
$z = 4(x')^2 + 2(y')^2$,
Elliptic paraboloid
$\theta = \tan^{-1}1 = 45^\circ$

26. $P = \frac{1}{\sqrt{2}}\begin{bmatrix} 1 & -1 \\ 1 & 1 \end{bmatrix}$
$z = 4(x')^2 - (y')^2$,
Hyperbolic paraboloid
$\theta = \tan^{-1}1 = 45^\circ$

27. $P = \frac{1}{\sqrt{10}}\begin{bmatrix} 3 & -1 \\ 1 & 3 \end{bmatrix}$
$z = 5(x')^2 - 5(y')^2$,
Hyperbolic paraboloid
$\theta = \tan^{-1}(1/3) \approx 18.43^\circ$

28. $P = \frac{1}{\sqrt{5}}\begin{bmatrix} 2 & -1 \\ 1 & 2 \end{bmatrix}$
$z = 10(x')^2 + 5(y')^2$,
Elliptic paraboloid
$\theta = \tan^{-1}(1/2) \approx 26.57^\circ$

29. $P = \frac{1}{\sqrt{(17)}} \begin{bmatrix} 4 & -1 \\ 1 & 4 \end{bmatrix}$ $z = 34(x')^2 + 17(y')^2$,
Elliptic paraboloid
$\theta = \tan^{-1}(1/4) \approx 14.04^\circ$

30. $P = \frac{1}{\sqrt{(13)}} \begin{bmatrix} 3 & -2 \\ 2 & 3 \end{bmatrix}$ $z = 52(x')^2 + 13(y')^2$,
Elliptic paraboloid
$\theta = \tan^{-1}(2/3) \approx 33.69^\circ$

In each of Problems 31-40 we list the critical point(s) of the given function, the values of

$$\Delta_1 = f_{xx} \quad \text{and} \quad \Delta_2 = f_{xx}f_{yy} - (f_{xy})^2$$

at each critical point, and its resulting classification.

31. Critical point $(-1,2)$: $\Delta_1 = 4$, $\Delta_2 = 8$
Local minimum

32. Critical point $(2,-3)$: $\Delta_1 = -6$, $\Delta_2 = 24$
Local maximum

33. Critical point $(-1/2,-1/2)$: $\Delta_1 = 4$, $\Delta_2 = -24$
Saddle point

34. Critical point $(2,-3)$: $\Delta_1 = 0$, $\Delta_2 = -1$
Saddle point

35. Critical point $(-3,4)$: $\Delta_1 = 4$, $\Delta_2 = 4$
Local minimum

36. Critical point $(6,-4)$: $\Delta_1 = 2$, $\Delta_2 = -8$
Saddle point

37. The equations $f_x = 3x^2 + 3y = 0$ and $f_y = 3y^2 + 3x = 0$ have the two solutions $(0,0)$ and $(-1,-1)$.
Critical point $(0,0)$: $\Delta_1 = 0$, $\Delta_2 = -9$: Saddle point
Critical point $(-1,-1)$: $\Delta_1 = -6$, $\Delta_2 = 27$: Local maximum

38. The equations $f_x = 2x - 2y = 0$ and $f_y = 3y^2 - 2x - 1 = 0$ have the two solutions $(1,1)$ and $(-1/3,-1/3)$.
Critical point $(1,1)$: $\Delta_1 = 2$, $\Delta_2 = 8$: Local minimum
Critical point $(-1/3,-1/3)$: $\Delta_1 = 2$, $\Delta_2 = -8$: Saddle point

39. The equations $f_x = 6 - 3x^2 = 0$ and $f_y = -3y^2 = 0$ have the two solutions $(\pm\sqrt{2},0)$. Because $\Delta_2 = 0$ at each of these critical points, Theorem 5 does not apply. However, a glance at the y^3 term in $f(x,y)$ makes it clear that there is no local extremum at either critical point.

40. The equations $f_x = 3y - 3x^2 = 0$ and $f_y = 3x - 3y^2 = 0$ have the two solutions $(0,0)$ and $(1,1)$.
Critical point $(0,0)$: $\Delta_1 = 0$, $\Delta_2 = -9$: Saddle point
Critical point $(1,1)$: $\Delta_1 = -6$, $\Delta_2 = 27$: Local maximum

SECTION 8.3

DIAGONALIZATION AND DIFFERENTIAL EQUATIONS

The coefficient matrices in Problems 1-8 in this section have already been diagonalized in Problems 1-8 in Section 6.2. Therefore we present here only the answers.

1. $x_1 = 2c_1e^{3t} + c_2e^t$, $\quad x_2 = c_1e^{3t} + c_2e^t$

2. $x_1 = 3c_1e^{2t} + c_2$, $\quad x_2 = 2c_1e^{2t} + c_2$

3. $x_1 = 3c_1e^{3t} + c_2e^{2t}$, $\quad x_2 = 2c_1e^{3t} + c_2e^{2t}$

4. $x_1 = 4c_1e^{2t} + c_2e^t$, $\quad x_2 = 3c_1e^{2t} + c_2e^t$

5. $x_1 = 4c_1e^{3t} + c_2e^t$, $\quad x_2 = 3c_1e^{3t} + c_2e^t$

6. $x_1 = 3c_1e^{2t} + 2c_2e^t$, $\quad x_2 = 4c_1e^{2t} + 3c_2e^t$

7. $x_1 = 5c_1e^{2t} + 2c_2e^t,$ $x_2 = 2c_1e^{2t} + c_2e^t$

8. $x_1 = 5c_1e^{2t} + 3c_2e^t,$ $x_2 = 3c_1e^{2t} + 2c_2e^t$

The coefficient matrices in Problems 9-12 here have already been diagonalized in Problems 1-4 in Section 6.4.

9. $x_1 = c_1e^{4t} + c_2e^{2t},$ $x_2 = c_1e^{4t} - c_2e^{2t}$

10. $x_1 = c_1e^{6t} + c_2e^{2t},$ $x_2 = 2c_1e^{6t} - c_2e^{2t}$

11. $2x_1 = 2c_1e^{10t} - c_2e^{5t},$ $x_2 = c_1e^{10t} + 2c_2e^{5t}$

12. $2x_1 = 2c_1e^{15t} - c_2e^{5t},$ $x_2 = c_1e^{15t} + 2c_2e^{5t}$

The coefficient matrices in Problems 13-16 here have already been diagonalized in Problems 11-14 in Section 6.4.

13. $x_1 = c_1e^{9t} + c_2e^{6t} + c_3$
$x_2 = c_1e^{9t} - 2c_2e^{6t}$
$x_3 = c_1e^{9t} + c_2e^{6t} - c_3$

14. $x_1 = c_1e^{9t} + 4c_3$
$x_2 = 2c_1e^{9t} + c_2e^{6t} - c_3$
$x_3 = 2c_1e^{9t} - c_2e^{6t} - c_3$

15. $x_1 = c_1e^{6t} + c_2e^{3t} + c_3e^{3t}$
$x_2 = c_1e^{6t} - 2c_2e^{3t}$
$x_3 = c_1e^{6t} + c_2e^{3t} - c_3e^{3t}$

16. $x_1 = c_1e^{9t} + c_2e^{6t} + c_3e^{2t}$

$x_2 = c_1e^{9t} - 2c_2e^{6t}$

$x_3 = c_1e^{9t} + c_2e^{6t} - c_3e^{2t}$

17. $x_1 = 6c_1 + 3c_2e^{t} + 2c_3e^{-t}$

$x_2 = 2c_1 + c_2e^{t} + c_3e^{-t}$

$x_3 = 5c_1 + 2c_2e^{t} + c_3e^{-t}$

18. $x_1 = c_2e^{t} + c_3e^{3t}$

$x_2 = c_1e^{-2t} - c_2e^{t} - c_3e^{3t}$

$x_3 = -c_1e^{-2t} + c_3e^{3t}$

19. $x_1 = c_1e^{2t} + c_3e^{3t}$

$x_2 = -c_1e^{2t} + c_2e^{-2t} - c_3e^{3t}$

$x_3 = - c_2e^{-2t} + c_3e^{3t}$

20. $x_1 = c_1e^{-t} + c_2e^{2t}$

$x_2 = -c_1e^{-t} + c_3e^{2t}$

$x_3 = - c_3e^{2t}$

In each of Problems 21-28 we list both the coefficient matrix A for the equivalent system $\bar{x}' = A\bar{x}$ and the solution $x(t)$ of the original differential equation.

21. $A = \begin{bmatrix} 0 & 1 \\ -2 & 3 \end{bmatrix}$ $\qquad x = c_1e^{2t} + c_2e^{t}$

22. $A = \begin{bmatrix} 0 & 1 \\ 3 & 2 \end{bmatrix}$ $\qquad x = c_1e^{3t} + c_2e^{-t}$

23. $A = \begin{bmatrix} 0 & 1 \\ 25 & 0 \end{bmatrix}$ $\qquad x = c_1e^{5t} + c_2e^{-5t}$

24. $A = \begin{bmatrix} 0 & 1 \\ -15 & -8 \end{bmatrix}$ $\qquad x = c_1e^{-3t} + c_2e^{-5t}$

25. $A = \begin{bmatrix} 0 & 1 & 0 \\ 0 & 0 & 1 \\ 0 & 9 & 0 \end{bmatrix}$ $\qquad x = c_1 + c_2e^{3t} + c_3e^{-3t}$

26. $A = \begin{bmatrix} 0 & 1 & 0 \\ 0 & 0 & 1 \\ 0 & -2 & -3 \end{bmatrix}$ $\qquad x = c_1 + c_2e^{-t} + c_3e^{-2t}$

27. $A = \begin{bmatrix} 0 & 1 & 0 \\ 0 & 0 & 1 \\ -2 & 1 & 2 \end{bmatrix}$ $\qquad x = c_1e^{t} + c_2e^{-t} + c_3e^{2t}$

28. $A = \begin{bmatrix} 0 & 1 & 0 \\ 0 & 0 & 1 \\ -4 & 4 & 1 \end{bmatrix}$ $\qquad x = c_1e^{t} + c_2e^{2t} + c_3e^{-2t}$

30. A simple computation yields

$$|A - \lambda I| = \begin{bmatrix} -\lambda & 1 & 0 \\ 0 & -\lambda & 1 \\ -r & -q & -p - \lambda \end{bmatrix}$$

$$= -(\lambda^3 + p\lambda^2 + q\lambda + r).$$

The equation

$$\begin{bmatrix} -\lambda & 1 & 0 \\ 0 & -\lambda & 1 \\ -r & -q & -p - \lambda \end{bmatrix} \begin{bmatrix} u \\ v \\ w \end{bmatrix} = \begin{bmatrix} 0 \\ 0 \\ 0 \end{bmatrix}$$

defines the eigenvectors associated with the eigenvalues $\lambda = \lambda_1, \lambda_2, \lambda_3$. The first two scalar equations are

$$-\lambda u + v = 0 \quad \text{and} \quad -\lambda v + w = 0,$$

so

$$v = \lambda u \quad \text{and} \quad w = \lambda v = \lambda^2 u.$$

Taking $u = 1$, we therefore get the eigenvector $(1, \lambda, \lambda^2)$ associated with the eigenvalue λ.

CHAPTER 9

NUMERICAL METHODS

SECTION 9.1

ROUNDOFF ERRORS, PIVOTING, AND CONDITIONING

The answers to Problems 1-8 and Problems 9-16 are arranged in double columns below, to facilitate comparison between the exact solutions on the left and the approximate solutions on the right.

1. $x = 3,\ y = 2$	9. $x = 3.00,\ y = 2.00$
2. $x = 5,\ y = -3$	10. $x = 4.99,\ y = -2.99$
3. $x = -4,\ y = 3$	11. $x = -3.91,\ y = 2.94$
4. $x = 5,\ y = 4$	12. $x = 4.99,\ y = 3.99$
5. $x = 4,\ y = -1,\ z = 3$	13. $x = 3.94,\ y = -0.975,\ z = 2.95$
6. $x = 3,\ y = 1,\ z = -2$	14. $x = 3.00,\ y = 1.00,\ z = -2.00$
7. $x = 1,\ y = 3,\ z = -4$	15. $x = 0.62,\ y = 3.08,\ z = -3.94$
8. $x = 1,\ y = 3,\ z = 5$	16. $x = 0.97,\ y = 3.01,\ z = 5.00$

Some of the solutions to Problems 9-16 exhibit the effects of a loss of significant digits that can occur when two numbers with the same number of signigicant digits are subtracted. For instance, in Problem 11 the Gaussian elimination with pivoting proceeds as follows:

$$\begin{bmatrix} 2 & 3 & 1 \\ 3 & 5 & 3 \end{bmatrix} \longrightarrow \begin{bmatrix} 3 & 5 & 3 \\ 2 & 3 & 1 \end{bmatrix}$$

$$\longrightarrow \begin{bmatrix} 1 & 1.67 & 1 \\ 2 & 3 & 1 \end{bmatrix}$$

$$\longrightarrow \begin{bmatrix} 1 & 1.67 & 1 \\ 0 & -0.34 & -1 \end{bmatrix}$$

The -0.34 in the second row is only a 2-digit approximation to the exact value -1/3 (and -0.33 would be better), resulting in

$$y = 1/0.34 = 2.94, \qquad x = 1 - (1.67)(2.94) = -3.91.$$

Note the very poor approximation for x in Problem 15, despite the fact that the results for y and z are not so bad. It rsults from back-substitution in the equation

$$x + 2.33y + 3z = -4,$$

yielding

$$\begin{aligned} x &= -4 - (2.33)(3.08) - (3)(-3.94) \\ &= -4 - 7.18 + 11.8 \\ x &= 0.62 \end{aligned}$$

17. $\lambda_1 \approx -0.00003, \qquad \lambda_2 \approx 3.33333, \qquad c(A) \approx 111111$

18. $\lambda_1 \approx 0.800, \qquad \lambda_2 \approx 500.2, \qquad c(A) \approx 625.5$

19. $\lambda_1 \approx 0.00008, \qquad \lambda_2 \approx 2.50002, \qquad c(A) \approx 31250$

20. $\lambda_1 \approx 0.005, \qquad \lambda_2 \approx 200.005, \qquad c(A) \approx 40000$

21. (a) $x = 1.00010001\cdots, \qquad y = 0.99989998\cdots$
    (b) $x = 0, \quad y = 1$ (without pivoting)
    (c) $x = 1, \quad y = 1$ (with pivoting)

22. (a) $x = 1/3, \quad y = 1, \quad z = -3/4$ (exact solution)
    (b) $x = 0.34, \quad y = 0.99, \quad z = -0.749$ (without pivoting)

(c) $x = 0.334$, $y = 1.00$, $z = -0.751$ (with pivoting)

23. $x = 1$, $y = 0$ with $\bar{b} = (2,1)$
$x = 0$, $y = 2$ with $\bar{b} = (2,1.0002)$

24. $x = 10$, $y = 0$ with $\bar{b} = (30,10)$
$x = 20$, $y = -30$ with $\bar{b} = (30,10.001)$

25. The solution before rounding is $x = 1000$, $y = -998$.
The solution after rounding is $x = 900$, $y = -898$.

26. The solution before rounding is $x = 500$, $y = 1008$.
The solution after rounding is $x = 300$, $y = 602$.

27. The length of x is given by

$$|x|^2 = x \cdot x$$

$$= (c_1v_1 + \cdots + c_nv_n) \cdot (c_1v_1 + \cdots + c_nv_n)$$

$$|x|^2 = c_1^2 + c_2^2 + \cdots + c_n^2$$

because the eigenvectors $v_1, v_2, \cdots, v_n$ are orthonormal. Since

$$Ax = c_1Av_1 + c_2Av_2 + \cdots + c_nAv_n$$

$$= c_1\lambda_1v_1 + c_2\lambda_2v_2 + \cdots + c_n\lambda_nv_n,$$

the length of Ax is given by

$$|Ax|^2 = (c_1\lambda_1v_1 + \cdots + c_n\lambda_nv_n) \cdot (c_1\lambda_1v_1 + \cdots + c_n\lambda_nv_n)$$

$$= c_1^2\lambda_1^2 + c_2^2\lambda_2^2 + \cdots + c_n^2\lambda_n^2.$$

Finally, the fact that $\lambda_1^2 \leq \lambda_i^2 \leq \lambda_n^2$ for each $i = 1,2,\cdots,n$ now implies that

$$\lambda_1^2|x|^2 \leq |Ax|^2 \leq \lambda_n^2|x|^2.$$

SECTION 9.2

ITERATIVE METHODS OF SOLUTION

For each of Problems 1-10 we list only the exact solution. In each it is clear after several iterations that convergence to this solution is occurring.

1. $x_1 = 2$, $x_2 = 1$
2. $x_1 = 2$, $x_2 = -3$
3. $x_1 = 3$, $x_2 = 2$
4. $x_1 = 4$, $x_2 = 3$
5. $x_1 = 4$, $x_2 = -1$
6. $x_1 = 2$, $x_2 = -3$
7. $x_1 = 3$, $x_2 = 1$, $x_3 = 2$
8. $x_1 = 2$, $x_2 = -1$, $x_3 = 3$
9. $x_1 = 2$, $x_2 = 1$, $x_3 = 1$
10. $x_1 = 2$, $x_2 = 2$, $x_3 = 1$

The exact solutions for Problems 11-20 are the same as those for Problems 1-10, respectively, and the convergence is faster with Gauss-Seidel iterations than with Jacobi iterations.

21. (a) With the system as given the Jacobi iteration equations are

$$x_1 = 101 + 100x_2, \qquad x_2 = 99 - 100x_1.$$

Starting with $x_1 = x_2 = 0$, we get $x_1 = 101$ and $x_2 = 99$ after the first iteration, $x_1 = 10001$ and $x_2 = -10001$ after the second iteration, and so forth, with the situation obviously deteriorating very rapidly.

(b) If the original equations are interchanged then the Jacobi iteration equations are

$$x_1 = 0.99 - 0.01x_2 \qquad x_2 = -1.01 + 0.01x_1.$$

Starting with $x_1 = x_2 = 0$, we now get $x_1 = 0.99$ and $x_2 = -1.01$ after the first iteration, $x_1 = 1.0001$ and $x_2 = -1.0001$ after the second iteration. Thus rapid convergence to the exact solution $x_1 = 1$, $x_2 = -1$ is occurring.

22. Suppose, for instance, that we are dealing with a 4×4 system and have already "fixed" it so that $a_{11} \neq 0$ and $a_{22} \neq 0$. If we cannot now make $a_{33} \neq 0$ either by interchanging the third and fourth equations or by interchanging the third and fourth variables, then the system must look like

$$\begin{aligned}
a_{11}x_1 + a_{12}x_2 + a_{13}x_3 + a_{14}x_4 &= b_1 \\
a_{21}x_1 + a_{22}x_2 + a_{23}x_3 + a_{24}x_4 &= b_2 \\
a_{31}x_1 + a_{32}x_2 \qquad\qquad\qquad\quad &= b_3 \\
a_{41}x_1 + a_{42}x_2 \qquad\qquad\qquad\quad &= b_4.
\end{aligned}$$

But then the lower left zero block contradicts the hypothesis that $|A| \neq 0$.

23. (a) First we find that

$$\begin{aligned}
r^{(k+1)} &= x^{(k+1)} - x^* \\
&= [Tx^{(k)} + c] - [Tx^* + c] \\
&= T[x^{(k)} - x^*] \\
r^{(k+1)} &= Tr^{(k)}.
\end{aligned}$$

Then $r^{(1)} = Tr^{(0)}$, $r^{(2)} = Tr^{(1)} = T^2r^{(0)}$, $r^{(3)} = Tr^{(2)} = T^3r^{(0)}$, and so forth.

(b) This follows because

$$T^k = (PDP^{-1})(PDP^{-1})\cdots(PDP^{-1}) = PD^kP^{-1},$$

where D^k is the diagonal matrix with diagonal elements $\lambda_1^k, \lambda_2^k, \ldots, \lambda_n^k$. Then, since each $|\lambda_i| < 1$, we see that $T^k \longrightarrow 0$ as $k \longrightarrow \infty$.

24. Simply take $f(x) = -20x^3$ in line 220, $g(x) = x^5 - 16x$ in line 230, and $h = 2/(n + 1)$ in line 250. The point is that $g''(x) = 20x^3 = -f(x)$ and $g(0) = g(2) = 0$.

25. When we replace $y'(x_i)$ by the difference quotient of Equation (9) in the text, the differential equation

$$y''(x_i) + y(x_i) = 0$$

yields the difference equation

$$\frac{y_{i+1} - 2y_i + y_{i-1}}{h^2} + y_1 = 0,$$

or

$$y_{i+1} - 2y_i + y_{i-1} + h^2 y_i = 0,$$

The appropriately altered version of Program TRIDIAG reads as follows:

```
100 REM--Program TRIDIAGB
110 REM--Finite differences solution of the diff
120 REM--equation  y" + y = 0  with initial
125 REM--conditions  y(0) = 0 and  y(PI/2) = 1.
130 REM
200 REM--Initialization:
210 REM
220      PI = 3.141593
230      DEF FNG(X) = SIN(X)
240      N = 9:   'N + 1 = no of subintervals
250      H = (PI/2)/(N + 1)
260      DIM Y(N)
270      FOR I = 1 TO N : Y(I) = 0
290      NEXT I
295 REM
300 REM--Successive corrections:
305 REM
310      Y(1) = (H*H*Y(1) + Y(2))/2
```

```
320       FOR I = 2 TO N - 1
330           X = I*H
340           Y(I)=(H*H*Y(I) + Y(I-1) + Y(I+1))/2
350           IF I/2 = I\2 THEN PRINT X, Y(I), FNG(X)
360       NEXT I
370       Y(N) = (1 + H*H*Y(N) + Y(N-1))/2
380 REM
400       GOTO 300  'For another iteration
410 REM--Press Ctrl/Break to stop execution
500       END
```

26.
$$\begin{bmatrix} 4 & -1 & -1 & 0 \\ -1 & 4 & 0 & -1 \\ -1 & 0 & 4 & -1 \\ 0 & -1 & -1 & 4 \end{bmatrix} \begin{bmatrix} u_1 \\ u_2 \\ u_3 \\ u_4 \end{bmatrix} = \begin{bmatrix} 0 \\ 1 \\ 0 \\ 1 \end{bmatrix}$$

27.
$$\begin{bmatrix} 4 & -1 & 0 & 0 & 0 & 0 \\ -1 & 4 & -1 & -1 & 0 & 0 \\ 0 & -1 & 4 & 0 & -1 & 0 \\ 0 & -1 & 0 & 4 & -1 & 0 \\ 0 & 0 & -1 & -1 & 4 & -1 \\ 0 & 0 & 0 & 0 & -1 & 4 \end{bmatrix} \begin{bmatrix} u_1 \\ u_2 \\ u_3 \\ u_4 \\ u_5 \\ u_6 \end{bmatrix} = \begin{bmatrix} 1 \\ 0 \\ 1 \\ 0 \\ 0 \\ 1 \end{bmatrix}$$

28. Put the initial vertex in class 1, the three that are one unit away in class 2, the three that are 2 units away in class 3, and the opposite vertex in class 4. Let x_n be the probability of reaching the opposite vertex if the spider is at a vertex in class n; let y_n be the expected number of steps. Then we have the equations

$$x_1 = x_2$$
$$x_2 = (x_1 + 2x_2)/3$$
$$x_3 = (2x_2 + x_4)/3$$
$$x_4 = 1$$

with solution

$$x_1 = x_2 = x_3 = x_4 = 1;$$

and the equations

$$\begin{aligned} y_1 &= 1 + y_2 \\ y_2 &= 1 + (y_1 + 2y_3)/3 \\ y_3 &= 1 + (2y_2 + y_4)/3 \\ y_4 &= 0 \end{aligned}$$

with solution

$$y_1 = 10, \quad y_2 = 9, \quad y_3 = 7, \quad y_4 = 0.$$

29. With the same notation as in Problem 28 we have the equations

$$\begin{aligned} x_1 &= x_2 \\ x_2 &= (x_1 + 2x_3)/3 \\ x_3 &= (x_2 + x_3 + x_4)/3 \\ x_4 &= (x_3 + x_4 + x_5)/3 \\ x_5 &= (2x_4 + x_6)/3 \\ x_6 &= 1 \end{aligned}$$

with solution

$$x_1 = x_2 = \cdots = x_6 = 1;$$

and the equations

$$\begin{aligned} y_1 &= 1 + y_2 \\ y_2 &= 1 + (y_1 + 2y_3)/3 \\ y_3 &= 1 + (y_2 + y_3 + y_4)/3 \\ y_4 &= 1 + (y_3 + y_4 + y_5)/3 \\ y_5 &= 1 + (2y_4 + y_6)/3 \\ y_6 &= 0 \end{aligned}$$

with solution

$$y_1 = 35, \; y_2 = 34, \; y_3 = 32, \; y_4 = 27, \; y_5 = 19, \; y_6 = 0.$$

30. Let x_n be the probability that Peter wins all given that he has n coins; let y_n be the expected number of plays

remaining when Peter has n coins. Then we have the equations

$$x_0 = 0$$
$$x_n = (x_{n-1} + x_{n+1})/2, \quad 1 \le n \le 14$$
$$x_{15} = 1$$

with solution

$$x_n = n/15, \text{ so } x_{10} = 2/3;$$

and the equations

$$y_0 = 0$$
$$y_n = 1 + (y_{n-1} + y_{n+1})/2, \quad 1 \le n \le 14$$
$$y_{15} = 0$$

with solution

$$y_n = 15n - n^2, \text{ so } y_{10} = 50.$$

31. With the same notation as in the preceding problem, we have the equations

$$x_0 = 0$$
$$x_n = (3x_{n-1} + x_{n+1})/4, \quad 0 < n < 15$$
$$x_{15} = 1$$

with solution

$$x_n = (3^n - 1)/(3^{15} - 1),$$

so

$$x_{10} = 244/59393;$$

and the equations

$$y_0 = 0$$
$$y_n = 1 + (3y_{n-1} + y_{n+1})/4, \quad 0 < n < 15$$
$$y_{15} = 0$$

with solution

$$y_n = 2n - 30(3^n - 1)/(3^{15} - 1),$$

so

$$y_{10} = 1180540/59393 \approx 19.87675.$$

32. With the same notation, we have the equations

$$\begin{aligned} x_1 &= x_2 \\ x_2 &= (x_1 + 2x_2 + x_3)/4 \\ x_3 &= 1 \end{aligned}$$

with solution

$$x_1 = x_2 = x_3 = 1;$$

and the equations

$$\begin{aligned} y_1 &= 1 + y_2 \\ y_2 &= 1 + (y_1 + 2y_2 + y_3)/4 \\ y_3 &= 0 \end{aligned}$$

with solution

$$y_1 = 6, \quad y_2 = 5, \quad y_3 = 0.$$

33. (a) $x_n = \frac{1}{2} x_{n-1} + \frac{1}{2} x_{n+1}$, $n < 1000$; $x_{1000} = 1$.

(d) $0 \leq x_n \leq 1$, so $B = 0$: $x_n = 1$.

(e) $x_n = 1$.

34. (a) $x_n = \frac{18}{37} x_{n+1} + \frac{19}{37} x_{n-1}$, $1 < n < 200$;

$x_0 = 0$, $x_{200} = 1$.

(d) $x_n = [(19/18)^n - 1]/[(19/18)^{200} - 1]$.

(e) $x_2 \approx 0.00000229851$.